Geography of Texas

People, Places, Patterns

Erik Prout
Department of Geography
Texas A&M University

Kendall Hunt
publishing company

Cover images © Shutterstock, Inc.
Copyright © 2012 by Erik Prout

ISBN 978-0-7575-9741-1

Kendall/Hunt Publishing Company has the exclusive rights to reproduce this work,
to prepare derivative works from this work, to publicly distribute this work,
to publicly perform this work and to publicly display this work.

All rights reserved. No part of this publication may be reproduced,
stored in a retrieval system, or transmitted, in any form or by any
means, electronic, mechanical, photocopying, recording, or otherwise,
without the prior written permission of the copyright owner.

Printed in the United States of America
10 9 8 7 6 5 4 3 2 1

Contents

PART ONE Introduction to the Geography of Texas 1
    1 Defining Texas 7
    2 Regional Geography 27

PART TWO Cultural-Historical Geographies of Texas 41
    3 A Geographical Past 45
    4 Cultural Diversity 63
    5 Cultural Landscapes 83

PART THREE Physical Geographies of Texas 111
    6 Physical Processes 115
    7 Physiographic Regions 141
    8 Human–Environment Interactions 169

PART FOUR Modern Human Geographies of Texas 189
    9 Millions of Texans 193
    10 Social Structures 215
    11 Leisure and Play 239

PART Five Regional Geographies of Texas 265
    12 Regions of Texas 269

Further Readings 293
List of Maps 295
Index 299

FIGURE 1-1 Welcome Sign at Texas Border
US-62 at New Mexico border with Guadalupe Mountain Range in background, Culberson County. Photo by Author.

WELCOME TO TEXAS

Thanks to all-you-all who helped me write this textbook Students have engaged me with stories and photographs of their home towns, and in return I tried to match their enthusiasm for learning about Texas. One, Kate Haas, recently helped with the cartography. While this edition is called "First, "it really is a follow up to the "Preliminary" edition; we caught numerous mistakes and introduced some improvements. Hopefully, it is ready to be used by universities other than Texas A&M. I acknowledge that there probably are errors and they are my own. I appreciate the professional interactions with Kendall/Hunt and their expertise of textbook production. Finally, I thank all the "Jaans" in life that make it all worthwhile .

Part One

Introduction to the Geography of Texas

Where is Texas? Not just a rhetorical question about direction or progress, but the start of a geographical inquiry. "Where is something located" is typically the first question a geographer asks about a place. Yet it is definitely not the only question. Geographers are interested in many different things including the people who inhabit and shape the Earth and all the variations of culture and ecology found around the world. Therefore, many questions come to mind. Who migrates to Texas? What trees grow in Texas? What languages do Texans speak? What crops do Texans plant? Why does the rest of the world think Texans ride horses to school?

We humans live on this planet, construct complex societies, and modify our physical environments. We also read and write, and we use those skills to build knowledge and increase understanding about who we are, where we are, and more generally what's going on around us. Now we have two complex topics: Texas and geography. Texas is our main topic, and it will take a complete semester to only scratch the surface. It is geography that is probably less known to most people yet it is the perspective of this instructor/writer. Geography or geographies (in the plural) are the preferred forms of knowledge and understanding used in this textbook.

Part One is organized into two distinct chapters: Texas and geography. Chapter One is an initial survey of and discussion about Texas. Basically, the first chapter starts our answer to the question "where is Texas?" Chapter Two is a condensed introduction to geography, so the reader can better understand the author. The second chapter includes the purpose of regional geography, some foundational geographical ideas, and a primer on maps, which is an important tool for geographers and an essential skill in geography.

LEARNING OBJECTIVES: GEOGRAPHY OF TEXAS INTRODUCTION

Each part of this textbook is associated with a specific course evaluation. Part One (Introduction) is testable material for every examination. Additionally, each part has a list of learning objectives, recitation goals, and important maps at the beginning before the individual chapters.

- Discuss the question—Where is Texas?
  In the form of an essay, could you discuss this question? An ideal essay would include latitude/longitude specifics, the difference between absolute and relative location, Texas' borders and neighbors, and appropriate use of geographical terms such as region.

- Define and elaborate the five geographical terms.

- Identify map types; use map reading skills.

- Differentiate types of regions.

Maps
Familiarity with general maps of Texas: neighbors and borders.
Reading general and academic maps; appreciating cartography.
Know where to find maps online and in library.

Definitions
Geography
Region: formal, functional, vernacular
Maps: topographic, thematic, isopleth, choropleth, cartogram
Location; Scale; Site and Situation
Latitude and Longitude

Web sites:
http://2010.census.gov
http://www.texas.gov
http://www.texasalmanac.com
http://www.windows.state.tx.us
http://www.lib.utexas.edu/maps
http://www.traveltex.com

Part One Introduction to the Geography of Texas 3

4   Part One   Introduction to the Geography of Texas

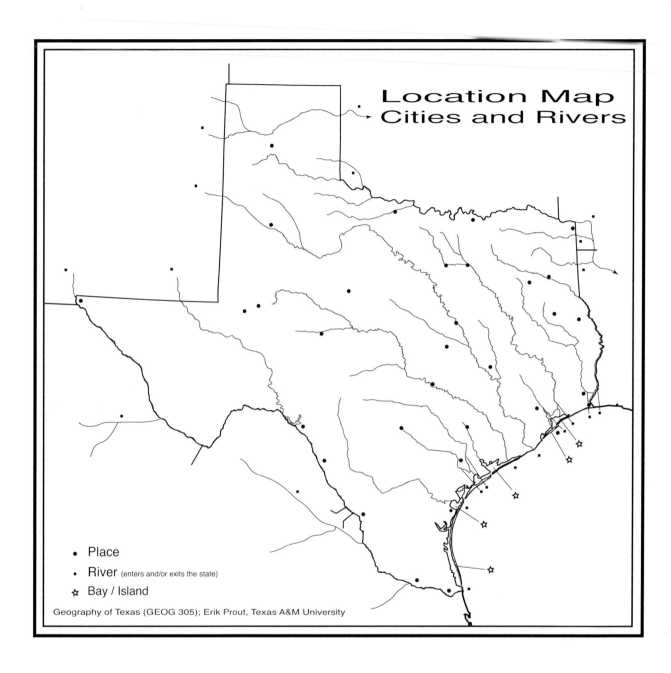

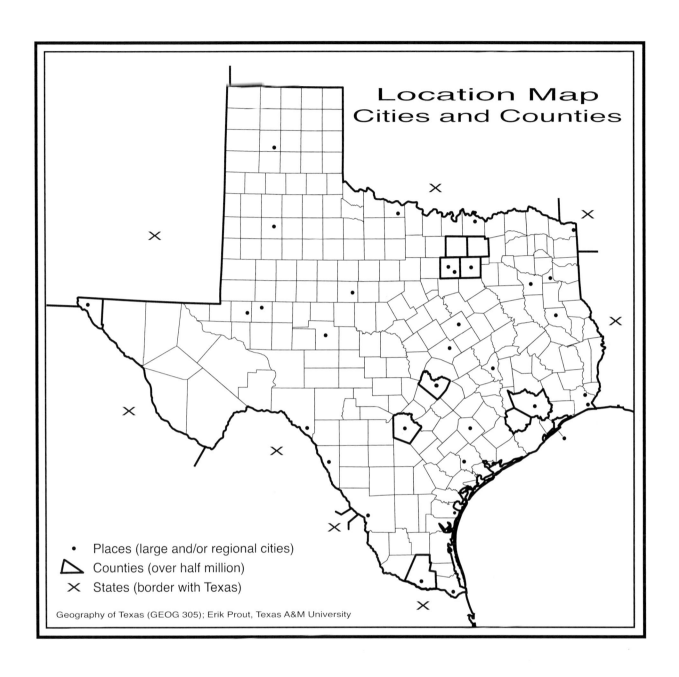

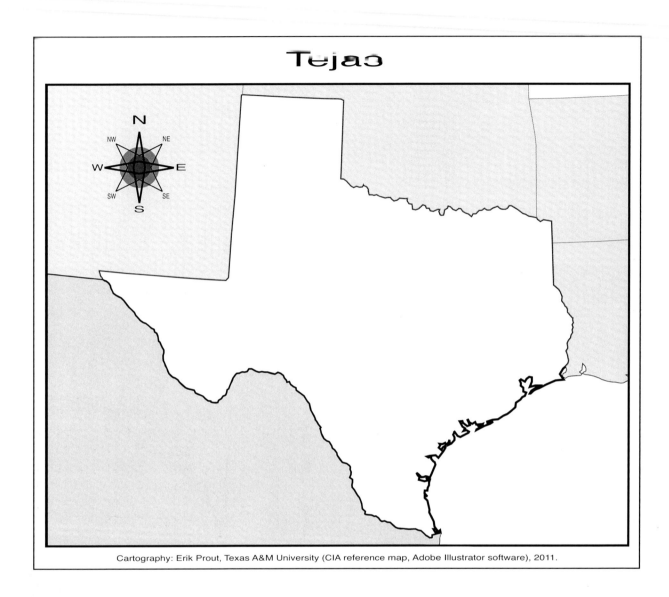

1
Defining Texas

Texas is big! We've all heard that expression in one form or another. No geographer would argue with Texas having a large territorial area when compared to the other American states. Yet we both know that the "big" is used in other ways such as boldness, spirit, ego, pride as well as a few negative connotations. Although it would be fun to identify geographies of big trucks, big belt buckles, big hair, and big personalities, we first need to articulate a basic answer to the where question—where is Texas? More to the geographic point, where is Texas located in the bigger picture of the Earth? In this chapter, we will first try to define both an essential and a locational answer to the where question; then, we will explore Texas' immediate cross-border neighbors and its one of a kind geographical shape.

ESSENTIAL DEFINITION

The essential definition of Texas is that Texas is a "political" entity. Like any other political entity, Texas has people and territory that are in it and distinct from other entities. Furthermore, polities have corresponding cultural and economical overlaps, but at the most essential, Texas is a state. A state is a political administrative unit defined in a democracy by its respective constitution, and Texas is legally defined by the U.S. and Texan Constitutions. Therefore, Texas is both singular as well as part of a larger political entity; its statehood is one of fifty that comprise the Unites States of America.

It is this comparison between Texas and the other 49 states that we use big accurately; Texas is both the second largest in area (after Alaska) and the second most populous (after California). While Texas definitely has unique qualities, it is most fundamentally American. I would argue that Texas is one of the crucial anchor points of American identity; it is not the sole anchor point, center, nor at one end of any spectrum. Basically, Texas is a known entity along with a few others: New York City, coastal California, and most importantly, the middle-American amalgamation. New York is the quintessential American city that defines the best and worst of urban America. The big apple, which was originally a derogatory reference to New York's apparent unequal share of resources, is the undisguised backdrop for comic books such as Superman and Batman. After 9/11, we all became New Yorkers in heart and spirit as it was understood to be an attack on America. On the opposite coast, Hollywood and the rise of modern cinema define California in the popular culture of practically the whole world. Californians construct themselves through direct democracy (anti-tax Proposition 13) and the occasional counter-culture movement (Hippies) that simultaneously attract and disgust other Americans. Texas may appear more "middle" American than the two coasts, but it differs from the normative of Iowa or Ohio. The Midwest with its manufacturing and agricultural synergy was a heartland often represented as all-American exuding typical America. In the big scheme of America, Texas exudes a boldness and patriotism which others can compare with themselves. Currently, it is scrutinized as part of the political base for the Republican Party symbolized by the Bush presidencies.

A state is also a type of region, which geographers utilize because of their *a priori* existence. Basically, Coahuila, Louisiana, and Texas come to us as already existing units—unlike the Cotton Belt and Border Zone, so there's no need to argue where the region begins and ends. Because they are states, these regions have a host of basic features such as measurable area, countable populations, and delineable borders. They also have economic and social interactions complexly operating at various scales including the regional level that coincides with their political boundaries. Examples include driver's licenses, educational standards, criminal codes, and tax collection that Texas and all states deal with in one form or another.

Moreover, states/regions have the three classic categories of geographical description: people, places, and patterns. People might be all inhabitants or only citizens (that can vote) and of course all the categories we derive from census like race and ethnicity. Places include everyday settings such as work, home, or school, or it might be that favorite picnic spot in a bustling city or tranquil countryside. Places can be both qualitative and location/scale specific, and on occasion, both those meanings of place may overlap, for example the State Capitol Complex (politicians and Austin) or Aggieland (Aggies and a University Campus). Patterns can be either human made such as the highway system or natural such as rainfall distribution. Most patterns can be depicted on maps, which mimic a look down perspective similar to what we experience in an airplane—except with abstract symbols. While the legal jurisdictions and administrative boundaries of states have well defined limits, such as the border between Texas and Chihuahua, Texas in the form of connotations may stretch beyond the territorial border. Texas in one's mind may even be boundless—especially with human imagination.

This boundless connotation is a constant theme concerning Texas, and it confounds any simple definition like statehood. The more obvious complication is that Texas was an independent political entity, a republic to be exact, which any school child in Texas can tell you. The imaginary memory of the Republic reinforces the "political" essence of my Texas definition, but it also introduces powerful cultural dimensions. While the Republic lasted for only a short period of time, the mythology of uniqueness and exceptionalism carries forth. Beyond this political independence, many Texans of that period envisioned a much larger entity that can be thought of as an empire. The notion of empire implies a unity and grandeur that subsumes all else—especially in terms of identity. Texas and Texan become icons in of themselves or a form of self-referencing symbol. The other obvious complication is territorial size. Because Texas is big, there is a logical need to sub-divide the state into manageable parts. We can identify not only both practical administrative needs but also intellectually to grasp its cultural and ecological diversity. Some obvious examples include district courts and state trooper stations, which deal with the public, but also regions like Panhandle or the Gulf Coast, which have different planning needs. The bigsize issue compels an author to choose between the entire Texas as a subject and the parts of the state with their regionalism and unique issues.

This dichotomy, if I can call it that, is just as fundamental as any other definition of Texas. Texas is both one and many! Terry Jordan opens his *Texas: a Geography* with just such a question: "Is it [Texas] one or many?" He recognized that many writers and scholars take a perspective of one or a perspective of many. The Texas-is-One group often perpetuated the empire mythology as well as over simplified if not overlooked the diversity of sub-regions in Texas. The Texas-is-Many group recognized the regional variations, which made it possible to overlook the powerful unifier of a grand Texas idea. We can reconstitute this dichotomy as *unity in diversity* (excuse the post-modern mantra), and ask ourselves which is more important. When unity is the bigger idea, it overwhelms the uniqueness of its components, and in terms of regions, the whole (Texas) is more important than the parts. When diversity is the bigger idea, the individuality of the components (places and sub-regions) overwhelms the complete composition. Perhaps, there is no winner, but a balance between perspectives is best. By necessity, this textbook uses both the one and many perspectives because there are moments when it's necessary to compare Texas with other states as well as highlight the variations within the state.

LOCATIONAL DEFINITION

The locational definition of Texas is perhaps less convoluted than the essential definition, but it is more technical. Location can be both exact (absolute) and relative (contextual to other locations), and only a geographer would venture to discuss location this way. First, I will discuss Texas around the language of location, and then I will state my best relative location statement in the conclusion. Fundamentally, Texas is located in the Northern and Western Hemispheres. The geographical center of Texas is 31° 08' North and 99° 20' West measured by the standardized latitude and longitude coordinates. The geographical center is identical in meaning as the average location; half of the state is on either side of their respective lines. The North (latitude) and West (longitude) in the coordinates reveal which hemisphere (or half of the Earth it is located in). What do these coordinates mean?

Being in the Northern Hemisphere means we have summer during the months of June through September and winter during the months of December through March. Summer means longer days (measured in hours of sunlight) and more intense, higher sun angles, which translates into the warmer half of the year. The Southern Hemisphere has the reverse seasonality: Christmas is one of the longest (and warmer) days in Australia and their shortest day is in June—exactly when we have our longest day. The dividing line between the Northern and Southern Hemispheres is the Equator, which is defined by the Earth–Sun relationship, and is often thought of as halfway between the polar points (North and South Poles). Texas is approximately one-third of the way between the Equator and North Pole, which means Texas is closer to the tropics than to the polar region. All of Texas is farther North than the Tropic of Cancer (23° 30'), so it would be incorrect to call Texas tropical. Texas is extremely farther South than the Arctic Circle (66° 30'), therefore it would be even more incorrect to think of Texas as so big to be polar. Most of Texas is accurately called mid-latitude (or lower mid-latitudes to be more precise), but as one goes southward, the term sub-tropical is proper. In many ways, Texas is in the transitional zone between sub-tropical and mid-latitude, which contributes to our unique and diverse environments.

Being in the Western Hemisphere has no direct physical meaning that being in the Northern Hemisphere does. East and West are relative to one another and often signify sunrise and sunset in ancient texts. In the case of longitude, we divide the Earth into 360 degrees, but there is no natural reference point. A prime meridian is necessary to reference everything ahead (East) and behind (West) of it as the Earth rotates. We do use longitude as a framework to create time zones with each time zone equaling approximately 15 degrees of longitude. Texas is six hours behind the current Prime Meridian (Greenwich) and, therefore, is classified as being in the Western Hemisphere. Another important consideration is that the Western Hemisphere is often called the "New World" or the "Americas" based on the events after Christopher Columbus' voyages. In scientific terminology, "Neo" is applied to life forms in the Americas and likewise "Paleo" to those from Asia, Africa, and Europe. The United States of America as its name implies is in the Americas.

The two largest landmasses in the Americas (Western Hemisphere) are defined typically as continents: North America and South America. North America extends from the Isthmus of Panama in the south to the Bering Strait in the northwest (separates Asia), which includes Greenland in the northeast (largest island in the world) and the Caribbean Islands (West Indies) in the Southeast. North America is

FIGURE 1-2 State of Texas

the third largest continent (after Asia and Africa) and has approximately 450 million inhabitants (fourth most people). Everything south of Panama is considered South America, which is smaller in area and has fewer people than North America. Obviously, Texas and the conterminous United States are located in North America. Technically, the United States formally includes territories and dependencies that are not contiguous, or even part of North America; for example, the State of Hawai'i and Guam in the Pacific as well as Puerto Rico in the Caribbean.

Texas is located on the eastern side of North America, which means it is on the Atlantic side of this large landmass. Despite the cultural-historical term Southwest, Texas rivers generally drain eastward or southeastward towards the Gulf of Mexico and not westward towards the Pacific. Because of the irregular shape of coastlines, it may not readily appear so but Texas has direct access to the Atlantic Ocean through the Gulf of Mexico. Therefore, Texas has open-sea access to the Atlantic realm that consists of East Coast cities like New York and Boston, near-by ports such as New Orleans, Havana, and Veracruz as well as places in Europe and Africa on the other side of the Atlantic Ocean.

Geographers who have tried to classify the regions of North America and/or the United States have not universally agreed on how to classify Texas. It exhibits a multitude of characteristics: South, West, Southwest, and even Midwest according to some. Texas is surely in the South because it was part of the Confederacy. The American South is defined still by the Civil War, but significantly, the plantation economy that entailed slavery and the legacy of race relations into the present. These Southern qualities are very pronounced in East Texas that borders other Confederate States.

Texas was also part of the westward moving American frontier, so it is also in the West. The open-range ranching and cowboy image establishes its western credentials and not just as a gateway or "Where the West begins" slogan in Fort Worth. The string of place-names with the word fort in it is a reminder of the Indian wars and difficult settlement of the continent's interior. The frontier mentality of self reliance and minimal government intrusion seems to persist long past the real actuality of vague borderless boundaries of settlement.

Southwest is another common term that once had a popular usage in college athletics when the Southwest Athletic Conference existed. Geographically, Southwest as a term is usually applied to the U.S.–Mexico borderlands, which Texas is a constituent part, and not the literal southwest corner of the country. The Southwest is a distinct American region because of its previous Spanish and Mexican history and the resulting cultural distinctions that vary from American standard. There is the lingering tri-cultural component of Indigenous, Hispanic, and Anglo groups as well as the impact of Catholicism. The region is a dynamic borderland between the United States and Mexico that is a contemporary gateway for migration and trade.

Midwest or Middle West is a term that has evolved as the western edge or frontier of America moved westward. Historically, Midwest was the area in between the older settled areas of the thirteen original colonies and the frontier. Currently, Midwest is West of the Appalachian Mountains and East of the Rocky Mountains, and should never be confused with the Far West (Pacific). Parts of Texas are on the Great Plains and exhibit some Midwestern qualities, but they don't dominate enough of the state to classify it that way. All four of these terms: South, West, Southwest, and Midwest have powerful qualitative meanings (formal culture regions in America), but they are not accurate in terms of geometric shapes or precise in terms of latitude and longitude.

One last locational term that is gaining popular usage is South Central, which tries to be exact but not in a formal way that implies cultural values. Recently agencies and corporations that use maps, such as the U. S. Weather Service and the Weather Channel, place Texas in the South Central category. Perhaps it is the overall shape of the conterminous 48 states that allows one to divide the area (or map of it) into six logical units using a north–south and a west–central–east divide. Therefore, Texas is in the southern half of the middle third (central) of the United States.

Summarizing the locational definition requires both an absolute and relative dimension. The absolute location can be determined from latitude and longitude lines (see Figure 1-1). Without implying any significance to the geometric shape, I'll relay the trivia because everyone eventually asks where are the special geographical spots in Texas? There are five distinct points that measure geographical center and the four extremities where one is farthest in one cardinal direction. There are also special border points, which will be mentioned with the borders discussion. The geographical center of Texas is located southwest of Mercury in McCulloch County. The latitude and longitude are 31° 08' N and 99° 20' W. Unlike the population center, which can constantly change, the geographical center is a geometric center. Because the borders of Texas are not likely to change, the site is for all practical purposes permanent. Unfortunately, the exact site is on private property, so the state has erected a roadside marker on a near-by highway. At the McCulloch County seat in Brady, residents have erected a monument on the county courthouse grounds, and the city of Brady advertises itself as the "Heart of Texas."

FIGURE 1-3 Heart of Texas Monument, Brady
McCulloch County Courthouse Square. Photo by Author.

The farthest East point in Texas is at 93° 31' W longitude. All locations in Texas have "West" longitudes, so the lower longitudes are in the East and higher longitudes are in the West. The farthest East point occurs along an eastward bend of the Sabine River in Newton County. The exact point is technically in the middle of the river downstream from Toledo Bend Reservoir. Both the farthest South and farthest West points fall along the Rio Grande. The southernmost point is found on a southward bend in Cameron County, which is southeast of Brownsville, very close to the river's end at the Gulf of Mexico. The latitude is 25° 50' N; the lower latitudes are more southerly for Texas because we are in the northern hemisphere. The farthest West point is along the Rio Grande in El Paso County at 106° 38' W longitude. The slight westward bend is between downtown El Paso and the New Mexico border and happens to be the small part of the Rio Grande that is a domestic border. The final point is not a point but a line of latitude. The farthest North in Texas is the Northern Panhandle border with Oklahoma at 36° 30' N, which includes five different counties.

Texas' relative location is best summarized as follows. Texas is located in the southwestern United States along the Mexican border. There are valid reasons to classify Texas as both South and West but especially Southwest, which makes it a most interesting crossroads of ideas as well as actual places. Fundamentally, Texas has a gateway to the Atlantic realm via the Gulf of Mexico, and it shares a long border with Mexico. It would be short-sighted to think of Texas as only a border province because it is not in the incipient national region or at the geometric center of the United States. Texas like pretty much everywhere has become interconnected to the national and international flow of ideas, peoples, and capital. Nevertheless, its proximity to Mexico is one of its most obvious relative factors; the broader Southwest is transforming itself in the new, open North America. Texas' future depends on how it acts and responds to Free trade, and the opportunities and challenges of its relative location.

BORDERS AND NEIGHBORS

This section explores the immediately adjacent regions to Texas and the dividing lines between them. Texas has eight states or if you will eight "neighbors" that are politically equivalent regions. Four of the eight neighbors are American states, and the remaining four are Mexican states. Louisiana and Nuevo León are our most populous neighbors; New Mexico and Chihuahua are our largest area neighbors. The eight states plus Texas are presented in Table 1-1 for comparison. Firstly, Texas has more residents than any of its neighbors and *almost equal* to all eight combined. Chihuahua is the largest Mexican state by area and Coahuila is the third largest. Yet none of Mexico's northern border states is heavily populated with between two and four million inhabitants respectively. Only Nuevo León exceeds the Mexican average for population density amongst Texas' Mexican neighbors. Texas' American neighbors are relatively smaller (except New Mexico) and less populated than average. Belying the image of Texas, the population density exceeds the American average.

Louisiana is only a remnant of its territorial extent before the Louisiana Purchase of 1803. Yet the Louisiana of today retains a French and Caribbean cultural imprint that makes it unique among American states. Arkansas was settled by Anglo-Americans just prior to Texas, so many Southern traits diffused through there slightly earlier. Oklahoma is just the opposite, it was originally designated a federal Indian preserve. Large-scale Anglo migration occurred much later (1890s) with the whole "Sooner" mythology. Finally, New Mexico belies its name except when compared to Mexico proper (Old Mexico). In fact the first European settlement in Texas was just a way station on the route to the Pueblo Indians in New Mexico.

14 Part One Introduction to the Geography of Texas

Northwest	North Central	Northeast
Southwest	**South Central** (Texas)	Southeast

Regional Schematic of United States

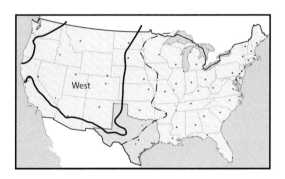
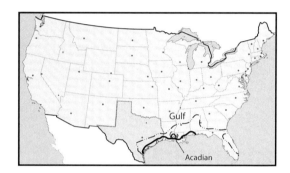

American Regions

Sources: Wilbur Zelinsky, 1980. North America's Vernacular Regions, Annals of the Association of American Geographers.

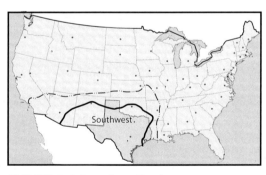
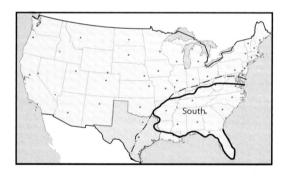

FIGURE 1-4 American Regions
Wilbur Zelinsky's research found that Texans perceived themselves to be in various American regions: South, Southwest, West, Midwest, and Gulf were most common Perceptual Regions.

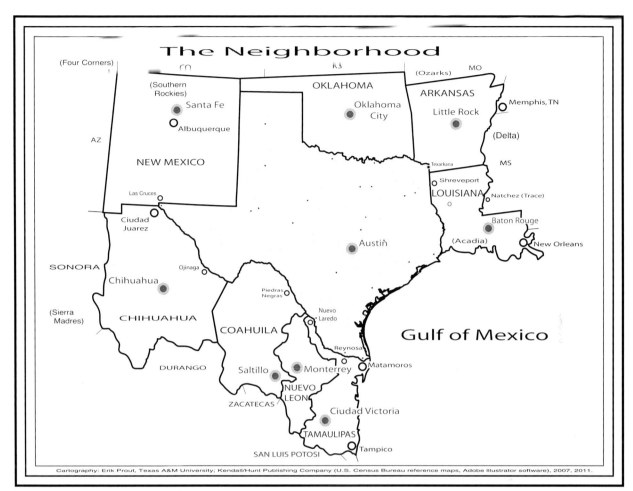

FIGURE 1-5 Texas Neighbors

TABLE 1-1 Neighboring States to Texas

State	Population (millions)	Pop. rank	Area (square km)	Area rank	Population Density (km)²	Pop.Den. PD rank
Louisiana	4.554	25/50	123,672	31/50	36.8	26/50
Arkansas	2.926	32/50	137,764	29/50	21.2	36/50
Oklahoma	3.765	28/50	181,188	20/50	20.8	37/50
New Mexico	2.067	36/50	314,927	5/50	6.6	47/50
TEXAS	**25.146**	**2/50**	**691,022**	**2/50**	**36.4**	**28/50**
Chihuahua	3.406	11/32	244,938	1/32	13.9	28/32
Coahuila	2.748	16/32	149,982	3/32	18.3	26/32
Nuevo León	4.653	8/32	64,924	13/32	71.7	14/32
Tamaulipas	3.269	13/32	79,384	7/32	41.2	22/32

Source: US Census Bureau. 2010

TABLE 1-2 Texas Comparisons

Rank	State	Population
1	California	37,254,000
2	TEXAS	25,146,000 (8.1% of US total)
3	New York	19,378,000
4	Florida	18,801,000
5	Illinois	12,419,000

Table Two (A)—Population (2000 Census)

Rank	State	Area (square miles)
1	Alaska	663,267
2	TEXAS	268,580 (7% of US total)
3	California	163,696
4	Montana	147,045
5	New Mexico	121,598

Table Two (B)—Area

Category	Ranking	
Number of Farms	1	247,500
Farm acreage	1	130,400,000 acres
Number of Businesses	3	522,336
Employment	2	9,231,955
Number of Immigrants	4	63,840

Table Two (C)—Other Rankings of Texas
Sources: Texas Almanac, US Census Bureau. Quick Facts

When one takes a serious look at these four states, it is possible to identify areas that are similar to regions of Texas. Typically, the cross-border dimension shows the most cultural and ecological similarities. For example, Cajun culture once extended into Southeast Texas, and likewise the Houston oil industry expanded into Southwest Louisiana. Therefore, we see similar surnames and economic activities on both sides of the Sabine Pass. Northeast Texas and Southwest Arkansas show similar cultural practices reflecting a common Southern heritage. North Texas and Oklahoma share oil production and cattle ranching as trademark businesses. Lastly, the Panhandle is comparable to the adjacent eastern counties of New Mexico, which are actually called the Texas counties. Basically, geographers explain that the West Texas culture area spread beyond the current Texas boundaries.

The four Mexican states also have similarities with their cross-border/cross-river components in Texas. While the length of the Texas-Mexico border is quite long, most people think of the entire region as South (or sometimes Southwest) Texas. It is just as common for them to lump all border states into a generic Mexico. Actually, Mexico has a federal system, like the United States, with 31 states and a federal territory (akin to the District of Columbia). Therefore, each Mexican state has its own history, regional government with elected governors, and (for the four that border the Rio Grande) unique issues with Texas.

South Texas carries its own Mexican-American and Mexican-Texan (Tejano) identity, so the obvious cross-border similarity is inherent. As J.B. Jackson wrote many years ago, rivers bring people together

TABLE 1-3 Border Segments of Texas

	Border Segment	Definition	Type	Neighbor	Character
I	Gulf of Mexico	0 Marine Leagues from tidal zone	Natural	None (U.S. territorial waters)	(Marine: federal and international dimension)
IIa	Rio Grande	Center of Stream	Natural	Tamaulipas	International
IIb	Rio Grande	Center of Stream	Natural	Nuevo León	International
IIc	Rio Grande	Center of Stream	Natural	Coahuila	International
IId	Rio Grande	Center of Stream	Natural	Chihuahua	International
IIe	Rio Grande	Center of Stream	Natural	New Mexico	Domestic
III	32° N Parallel	Line of Latitude	Geometric	New Mexico	Domestic
IV	W. Panhandle / 103° W Meridian	Line of Longitude (as surveyed)	Geometric	New Mexico	Domestic
V	N. Panhandle / 36° 30' N Parallel	Line of Latitude	Geometric	Oklahoma	Domestic
VI	E. Panhandle / 100° W Meridian	Line of Longitude	Geometric	Oklahoma	Domestic
VIIa	Red River	South/West (Right) Bank	Natural	Oklahoma	Domestic
VIIb	Red River	South/West (Right) Bank	Natural	Arkansas	Domestic
VIIIa	94° 04' W Meridian	Line of Longitude	Geometric	Arkansas	Domestic
VIIIb	94° 04' W Meridian	Line of Longitude	Geometric	Louisiana	Domestic
IX	Sabine River (Sabine Lake and Pass)	Center of Stream	Natural	Louisiana	Domestic

around a single resource. Jackson goes on to say that it is an accident of history for the Rio Grande to be used as a border when people on both sides were really one culture. For the residents of El Paso and Juarez, the Chihuahuan Desert besets an arid environment difficult for human settlement and simply put the river brings life. The Rio Grande is awkwardly both a unifier (water in the desert) and a divider (international border).

Chihuahua is the farthest inland state, which we associate with the Trans-Pecos Texas. Chihuahua actually shares a land border with New Mexico. This is the state J.B. Jackson wrote about in his article on the Rio Grande. Coahuila was the inland gateway to the central Plateau of Mexico for early Texas via its frontier capital of Saltillo (and Montclava temporarily). Texas historians know Coahuila as the Mexican state that Texas was adjoined to in the 1830s. Despite its very small Texas border, Nuevo León is the current, functional gateway. Its capital Monterrey is a bustling regional center. Finally, the state of Tamaulipas on the Gulf of Mexico shares oil resources Fishing with Texas along their adjacent coastal zones. Tamaulipas was previously called *Nuevo Santander* when Mexico claimed the Nueces River as their northern border.

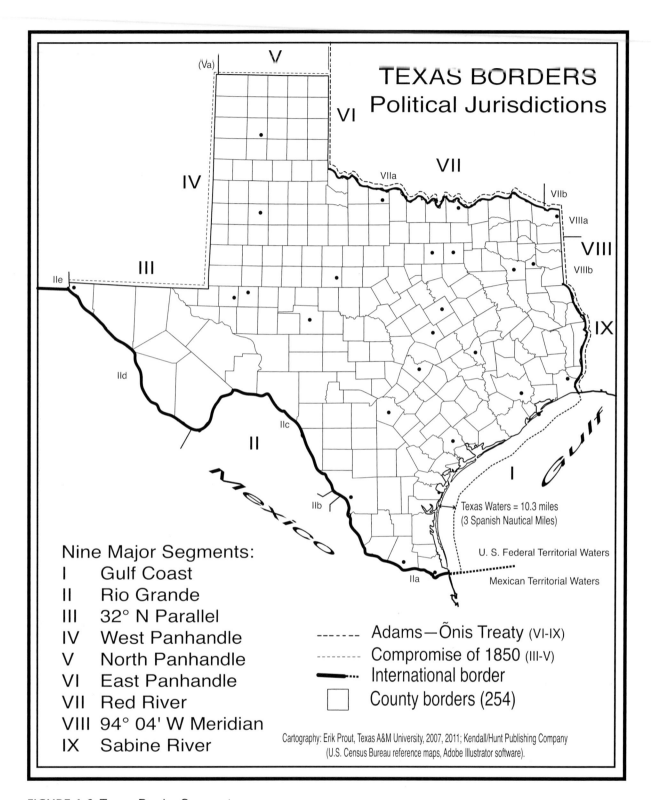

FIGURE 1-6 Texas Border Segments

TABLE 1-4 Length of Texas Borders

	Border Section	Length (standard arcs)	Length (with meanders/ tidal zone)	Major/Selected Crossing Points
I	Gulf Shoreline	367.0	624	Galveston (I-45) Corpus (I-37) Port Lavaca (US-87)
II	Rio Grande (international)	889.0	1254	**Controlled Access** I-35 (Laredo) US-77 (Brownsville) US-54 (El Paso)
III	Latitude 32° 00' N	209.0	same	I-10 (El Paso) US 285, 62, 180
IV	W. Panhandle 103° ?' W	310.2	same	I-40 (old US 66) US 60, 70, …
V	N. Panhandle 36° 30' N	167.0	same	US 385, 287, 54, 83
Vi	E. Panhandle 100° 0' W	133.6	same	I-40 (old US 66) US 60, 62
VII	Red River (SW bank)	480.0	726	I-35 I-44 (Wichita Falls) US 283, 70, 75, …
VIII	Longitude 94° 04' W	106.5	same	I-30; I-20 US 59, 71, 80, 79
IX	Sabine River / Lake / Pass	180.0	292	I-10 (Orange) US 84, 190, (old 90) OSR (TX 21)
Total	Texas Outline	2842.3	3822	

Source: Texas Almanac.

Table 1-2 presents some rankings for Texas. Part one shows the five most populous American states, and part two shows the five largest by area. The final part shows some other rankings of Texas compared with the 50 other American states. Next, we turn to Texas' borders (Table 1-3).

From the Texan perspective, there are nine border segments. Some of these segments can be divided further due to the changing context on the other side of the border. If we desired even more nuance, we could argue for close to twenty identifiable segments. The most important characteristic of all borders is how the two parties involved define the border. For example, the U.S. and Mexican governments use the Rio Grande as part of their border with each other. Although it seems intuitive to use natural features for boundary lines because they "draw themselves," there are inherent problems such as what to do if the river changes course. The U.S. and Mexico created a Boundary Commission back in 1884 to handle such issues. Since then it has been expanded to deal with important questions about water usage. Currently, the International Boundary and Water Commission meets in El Paso to address those issues. There are instances around the world when both parties do not agree on a border. In this case, the more powerful

FIGURE 1-7 Border on US-285, Reeves County
Texas–New Mexico border at 32°N Parallel. Photo by Author.

party typically dictates their claim, or occasionally, they agree to disagree with a neutral zone or cease-fire line with final boundary to be determined later.

The other common method of defining a border is to draw a straight line or some other geometric line such as an arc (around an island). Most straight lines utilize a meridian or parallel line because they map easily, and then the parties must delineate the border on the Earth's surface. Five of Texas' nine border segments are geometric using both lines of latitude (parallels) and lines of longitude (meridians) multiple times. Not all straight lines have to be exactly north-south or east-west, but they must have accurate end points. Many counties in Texas use such straight lines; even more apparent, counties formed after 1850 are typically rectangular.

An additional feature of borders is their character. In Texas, the difference between domestic and international borders is quite dramatic because of the nature of the U.S.–Mexican border. It's not just an international border, it's a line between the first world and third world with dramatic socio-economic differences. The border is partially militarized (or heavily policed) with fences, lights, and constant patrols to prevent unauthorized border crossings. In contrast, the U.S.–Canada border in Minnesota is very different and has only simple checkpoints. Texas' domestic borders like all domestic borders in the United States are completely open. Domestic borders inside the United States typically have only welcome signs demarcating and announcing jurisdictional changes.

Lastly, borders can be measured. Texas has almost three thousand miles of border. Less than one-third of the border is with Mexico, therefore, two-thirds is domestic. Likewise, approximately two-thirds of Texas border length consists of natural rivers, and the remaining one-third are geometric lines. Measuring the lines is relatively easy compared to natural features. There is a standardized way to measure curvy features such as coastlines and rivers that derives a relative figure. When using the standard arc method, Texas has 2,842 miles of border. When the actual coastline and river meanders are measured, Texas has closer to 3,822 miles of border (see Table 1-3 and Figure 1-6).

The first border segment differs from all the rest because it is a marine boundary with no obvious counterpart region on the other side. An obvious question to ask is where does Texas end and the Gulf begin? As most beach-goers know, the line is where sand becomes water. Unfortunately, the shoreline is

TABLE 1-5 Texas Border Posts

Border Post (segments)	Where is it?	How do I get there?
corner 1/2	Mouth of Rio Grande —Gulf Shoreline —Center of Stream	Swim/boat INS permission?
2a/2b 2b/2c 2c/2d 2d/2e	Rio Grande —Center of Stream (U.S.-Mex border) —Mexican domestic borders —d/e international to domestic	Swim/boat INS permission?
corner 2/3	Rio Grande 32° 00' N parallel	Swim/boat (see from I-10)
corner 3/4	32° 00' N parallel 103° ?' W meridian	FM 1218?
corner 4/5	36° 30' N parallel 103° ?' W meridian	US 56/64 (NM)
5a/5b?	Along 36° 30' N parallel —NM–OK border (103° 00' W meridian)	
corner 5/6	36° 30' N parallel 100° 00' W meridian	
corner 6/7	Red River (right bank of Prairie Dog Town Fork) 100° 00' W meridian	Walk/ride
7a/7b	Along the Red River —OK–AR border	Walk/ride
corner 7/8	Red River (right bank) 94° 04' W meridian	US 59?
8a/8b	Along the 94° 04' W meridian —AR–LA border	TX 77?
corner 8/9	Sabine River (center) at the 32° 00' N parallel —94° 04' W meridian	Swim/boat
corner 9/1	Sabine Pass —Gulf Shoreline —Center of Stream	Swim/Boat (see from TX-82)

constantly changing due to tides and storms; furthermore, how do we take into account inlets and islands. Despite using simple drafting techniques, maritime law is rather complex because of the layers of historical traditions. This fact of maritime status makes this border more complex than first imagined.

The "Gulf" border stretches from the mouth of the Rio Grande to the Sabine Pass. Starting at the outer edge of the tidal zone (area between average high and low tides), Texas' territorial waters extend out three Marine Leagues. A league varies around the world but it usually equals three nautical miles, which ironically varies in absolute length. Regardless, Texas territorial water extends to 10.36 miles. This differs from other U.S. states because three Marine Leagues was the Spanish standard back in 1836 and nobody clarified the claim until offshore oil drilling became profitable. Beyond three leagues, it becomes American territorial waters and eventually international waters (no claims are recognized except the US mexi Exclusive Economic Zone respective).

22   Part One   Introduction to the Geography of Texas

FIGURE 1-8 US-Mexico Border Crossing, Lower Rio Grande
Busy international border crossings for Texas are all along the Rio Grande; therefore, the only legal ports of entry are bridges and a few ferries. Photo courtesy of Texas Department of Transportation.

FIGURE 1-9 Geographic Center of Texas Marker
Geographic Center is near this park site on US-377 Southwest of Mercury, McCulloch County. Photo by Author.

The other eight borders are known as land borders even when some of them are technically rivers. As such they have boundary markers, which have exact locations. The markers are tangible icons that people can become fascinated with. Most people will see welcome signs when they cross the border on a major road or interstate highway. The markers where major border sections come together are often called corners: therefore, Texas has nine corners. In the next section, I'll describe both the segments and markers together; instead of being systematic (CW or CCW), let's follow a historical order.

The earliest boundaries and borders of/in Texas were of Spanish and Mexican origins; many internal and external borders have been retained up to the present. The Adams–Oñis Treaty (1819), for example, still defines four of Texas' border segments: sections six through nine or the easternmost ones. The Treaty was between the United States and Spain to define their common border after the ambiguity of the Louisiana Purchase. The border is described as follows:

> Starting at the Sabine Pass and Gulf of Mexico (Post 9/1), the border follows the center of stream of the Sabine Lake and Sabine River up to the 32° N parallel. At that point (Post 8/9), a line directly North to the Red River, which turned out to be the 94° 04' W meridian. At that point (Post 7/8), the border goes upstream to the 100° W meridian. At that point (Post 6/7), the border goes directly North along the 100th meridian to the Arkansas River.

The treaty went to describe the border all the way to the Pacific Ocean.

On the other end of the Gulf segment is the endpoint of Texas' longest border segment (Post 1/2). The Rio Grande border starts at its mouth and continues upstream in the center of the river for 889 miles. Along the

FIGURE 1-10 Border on US-87, Dallam County
Texas—New Mexico border at 103° W Meridian. Photo by Author.

way, the border passes all four Mexican neighbors and for a small section, New Mexico. This border was contentious and disputed between Texas and Mexico and later U.S. and Mexico. Texas initially claimed that the river to its source (starting point) was its South / West border eventually joining up with the Adams–Oñis lines at the 42nd parallel.

Finally, the last three segments and the all important corners (Posts 2/3 and 5/6) were defined in 1850. The Compromise of 1850 was between the federal and state governments to finalize the boundaries after Mexican–American War. Starting at the 100° W meridian (Eastern Panhandle), the border post (5/6) is at the 36° 30′ North. The Northern Panhandle border follows the 36 and a half parallel to the 103° W meridian (Post 4/5). The Western Panhandle border is usually stated as the 103° W meridian, but the actual border is slightly West of the meridian. The final segment (3) runs along the 32° N parallel from the Western Panhandle (Post 3/4) to the Rio Grande (Post 2/3).

Most border markers will never be seen unless one looks hard for them. Visiting the corners is quite difficult and it's even harder to get to the exact spots. Nevertheless, they are significant because they demarcate the limits of Texas, which is our region of study. Together, all the border segments create a unique shape.

SHAPE OF TEXAS

All geographical entities have a particular spatial shape. Basically, the outlines of things like states have a unique shape that we can learn to recognize. Many people learn their basic geography (U.S. states and countries of the world) by memorizing their name, their spot on the map, and the visual recognition of the shape. The shape is like a kind of jigsaw-puzzle piece; typically each state has a different color. Richard Francaviglia wrote a one of a kind book called *The Shape of Texas*. In his book, Francaviglia explored the

FIGURE 1-11 Shape of Texas Sign
Star Cafe combines Texas symbols: New Ulm, Austin County. Photo by Author.

FIGURE 1-12 Star Burger, Williamson County
State pride and fast food architecture; Texas State Highway 29 between Georgetown and Liberty Hill. Photo by Author.

use of Texas as a recognizable icon that was being used for marketing purposes as well as pride. If one thinks about it, the Texas flag, a lone star, and the state outline are all used for these reasons.

Nearly everyone in Texas already has a mental image of the state outline. Interestingly, the state outline is very conspicuous in the cultural landscape. In this chapter, we've explored the definitions of Texas from an abstract essential perspective as well as exacting and relative locational perspectives. We've also isolated and examined the border segments that make up the literal outline of the state. I conclude Chapter One with a few more photos that depict this "shape" theme. The cultural landscape is full of examples of Texas pride, and many of them can be looked at for geographical meaning. In the next chapter, we turn to academic geography and define region and explore the purpose of regional geography.

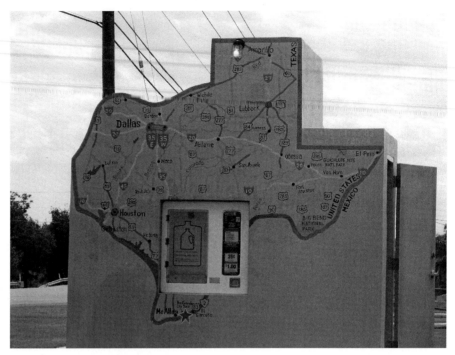

FIGURE 2-1 Watershop in the Shape and Map of Texas
Rio Grande City, Starr County. Photo by Author.

TEACHING TEXAS: IMPROMPTU LECTURE

Geography of Texas as a university course is increasing as the state's population grows. While a few examples of geographical scholarship and pedagogy exists, the first systematic textbook was published in 1984 (Terry Jordan's Texas: A Geography). If it was available, I'd assign it as essential reading. Instead, I read it thoroughly for my own lectures, and clearly was influenced by it as I wrote this textbook. The complexity of the state and the shear distances involved for exploring it in person ensure that it will always be a work in progress. Like all good things, the beauty of discovery never ends—just like the Highwaymen sang, "The Road goes on Forever, and the Party never Ends".

2
Regional Geography

Texas is a region. In fact, Texas is a very complex region, but what is a region? In Chapter Two, we turn our attention to academic geography and some important geographical ideas. Geography is one of the oldest educational pursuits, and there are numerous ancient texts that we rightfully call geography such as Strabo's *Orbis*. Strabo wrote elaborate descriptions of the Roman Empire detailing the nations (peoples) and provinces (regions) contained within it. Strabo did what people have always done—describe their worlds in as much detail as possible. Unfortunately, our list of ancient texts is very short and limited to the most recent millennia, yet we can assume some universal principles.

Geography in its most common form is space. Basic spatial cognition, such as orientation and discovery, is an integral part of being human. Meanwhile, drawing maps is an advanced form of language—also unique to humans. Imagine cavemen and cavewomen sitting around the campfire; someone tells young Grog to walk up to the ridgeline and return with a report of what he saw. Alas, we have prehistoric geography: simple description of the unknown. As one describes what's beyond the horizon, she or he is doing geography. Not all geography has to be real exploration, but bringing back direct observations and/or new perspectives is still at the heart of geographical inquiry.

Geography is both commonplace and learned in that one can be interested and engaged without formal training. A popular example is the *National Geographic* magazine and cable television station, which relies on spectacular photographic images. Commonplace geography includes not only *National Geographic* but also the everyday things we do such as driving to the university. We may not even be aware that we are thinking spatially when we know subconsciously where to make a turn on the way to university. Moving or navigating successfully, such as driving without getting lost, on the Earth's surface requires spatial reasoning. It is also very common to hear other places in the world mentioned during the news, and we process the information accordingly. For example, when we hear that American casualties occurred in Iraq as part of the conflict, we are sad but we don't have to change our daily routine. Unless we know someone in Iraq, we assign the events and tragedies as being far away. Yet a local stabbing or mugging is closer to home. Despite the significantly lower level of violence, we may reassess our personal security and take precautions to avoid perceived dangerous areas.

EARTH WRITING

On the other hand, geography can be an intellectual pursuit that searches for the philosophical meanings of space, place, and even human existence on the planet. There are many definitions of geography; here is a widely used one: geography is the study of the Earth as the home of humans. While clearly placing people at the center of attention, this definition also understands that the Earth is our essential habitat. Yet my favorite definition of geography is simply "earth writing." If one looks at the etymological roots of the word: "geo" is Greek for earth and "graphy" is translated as either writing or etching. Therefore, by choosing writing over other translations, earth writing is the literal definition of geography. Many people confuse geology with geography; the "-ology" is usually considered as study so geology is the study of the Earth.

Geographers quite literally write about the Earth. More specifically, geographers write about the people, places, and patterns that we find on the Earth. Our observations, whether they are scientific or humanistic, commonly refer to the human—environment component of the planet. We humans occupy the surface *(terra firma),* meanwhile, we avoid a majority of the Earth's surface (oceans). Because we prefer and thrive in specific environments, it's not surprising that much geographical research is about those places. The ways in which we write can vary as well: first, we describe; second, we try to explain if possible; third, if appropriate, we attempt to predict. All together, geographers describe, explain, and predict the people, places, and patterns on or near the Earth's surface where humans live and interact with their environments. Academic geographers, such as me, search for new and improved ways to observe the world around us and to incorporate this into the preexisting knowledge.

Like other academic disciplines, geographers tend to specialize. The three most important subcategories are human geography, physical geography, and geo-spatial techniques; in addition, many geographers have a regional specialty such as Europe or Africa. Human geography, sometimes called cultural geography, includes specializations in political, economic, and population to name just a few. Physical geography specializations are numerous and include geomorphology, climatology, hydrology, and biogeography. The techniques subcategory is changing dramatically as technology transforms our world, but cartography, remote sensing, and geographical information systems (GIS) are the main specializations. Geography as a discipline accommodates very different interests that collectively or even as individuals cross over categories such as natural sciences and social sciences.

Big Ideas

A few years ago, a distinguished group of geography scholars published a list of "big questions" in academic geography. The geographers were all leaders of the national association that represents a large number of academic and professional geographers in the United States: the Association of American Geographers (AAG). The AAG is the largest membership association of geographers allowing membership from anywhere; their annual meetings are overwhelmingly the largest gatherings of geographers in the world.

The AAG leaders were responding to a challenge about geography's relevance in the contemporary period. Their collective response may surprise you. Here are the 10 ideas posed as questions.

1. What makes places and landscapes different from one another and why is this important?
2. Is there a deeply held human need to organize space by creating arbitrary borders, boundaries, and districts?
3. How do we delineate space?
4. Why do people, resources, and ideas move?
5. How has the Earth been transformed by human action?
6. What role will virtual systems play in learning about the world?
7. How do we measure the unmeasurable?
8. What role has geographical skill played in the evolution of human civilization, and what role can it play in predicting the future?
9. How and why do sustainability and vulnerability change from place to place over time?
10. What is the nature of spatial thinking, reasoning, and abilities?

There are many ideas surrounding geography, and it should be apparent that educated people should be knowledgeable of these questions. The foundation for truly understanding these ideas is to learn the

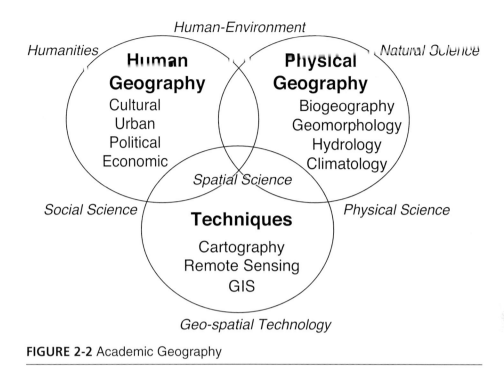

FIGURE 2-2 Academic Geography

basics. For pedagogical reasons, I ask my students to learn some key terms. The five basic terms to know are: geography, region, location, scale, and map. Some other important terms that I will discuss later include landscape and place. Geography has already been defined as Earth-writing, so let's turn to region.

REGIONS

Region was once the core idea of academic geography, and most geographers were obliged to be area specialists. While academic geography moved toward thematic specialization, regional geography remains a component of an undergraduate geography curriculum. Most non-geography majors will be exposed to a regional geography when they take a geography course. In Chapter One, I described an essential definition of Texas and its borders, which already tells us a lot about this region. Let's begin this section with a definition of the term.

A region is defined as a collection of places or a sizable area with common characteristics. Typically, regions consist of multiple known spots, more properly called places. Sometimes a large area has no known places, which ironically helps define it. Therefore, a region exists because we know its multiple parts or we sense an expansive amount of space. Moreover, we assign it a value based on some commonality. Those characteristics that are common throughout the places or area are typically how we define the region.

A region can be defined by a single feature such as a dominant crop, commonly spoken language, or obvious element of the natural landscape. Numerous examples of dominant crops include the Corn Belt and Cotton Belt, which implies both what one sees in the landscape as well as what the local inhabitants are doing. The term "belt" is not necessary but popularly used. There is also an American cultural region called the Bible Belt, therefore, a religious fervor or revitalization can be used to define a region. Sometimes a region relies on multiple characteristics instead of a single trait. For example, South Texas is both lower latitude with a subtropical climate and vegetation and a border zone adjacent to Mexico with a strong Hispanic cultural component. Most regions have a multiplicity of characteristics, so there's a nuance to defining and differentiating regions depending on what variables are most important to the person generalizing.

Regional Geography

Region is an extremely malleable concept that allows much imagination. The regional concept is important because it allows one to discuss parts of the whole or collective whole of the parts. For geographers, this means some spatial scale in between global and local. We can discuss the relative parts of the global whole or the aggregate whole of multiple local parts. Furthermore, one can use region to add and divide other intermediate levels such as U.S. states. For example, we can subdivide Texas into regions (e.g., East Texas) or we can add Texas with other states to create an American region (e.g., Southwest).

Regional geography is sometimes referred to as a thematic course. A geography of Texas course is about Texas—Texas being the theme. Just like the geography of Europe or Latin America course is somewhat self-defined; they would thematically be about Europe and Latin America respectively. A world regional geography course is about the whole world, but it is so broad that it can only survey the major world regions. Meanwhile, a course just on Texas (or California or the Great Lakes) can focus much more attention to the local details.

Regional geography courses can be both an inherent subject and an approach to teaching. What all regional geography courses have in common is that they mix human and physical geography. These courses try to focus on the region as opposed to the systematic specializations. When teaching world regional geography, it's even possible to discuss in a regional approach (one world region at a time). Yet when teaching regional geography of a state, it appears to be systematic because we take long looks at the state with each perspective.

Types of Regions

There are an infinite number of possible regions when we consider all the possible variables that one can choose from, and differing ways of encountering the world and of organizing ideas. Nevertheless, we can generalize most all regions into types or categories as well as identify some universal aspects that apply to most regions. Let's look to types of regions, which derives from cultural geography yet is applicable to most geographies, and then to the characteristics of regions.

There are three types of cultural regions: formal, functional, and vernacular. Formal regions are areas of the Earth with a coherent unifying characteristic. A previously mentioned example was Cotton Belt, which is a formal region. The Cotton Belt implies both a visual landscape associated with fields and processing gins as well as the socioeconomic activities of the people. Formal regions based on agriculture often correlate with a climatic variable such as temperature and/or precipitation that support the growth of specific plant life. They also tend to reveal a specialization and accumulation of knowledge and services associated with the economies of the agricultural product. Another broad example of a formal region is language, which can be simplified to what language is spoken most frequently on the Earth. A specific one is the Francophone region, or the part of the world where French is spoken by the majority of the people. In addition to France, Quebec (Canada), Southern Louisiana, and former French colonies around the world speak French as a first or second language. Formal language regions are apparent in the landscape because of the public use of the language on road signs, businesses, etc. Often formal regions are called uniform regions because of the uniformity of landscapes.

The second major type of cultural region is known as functional. A functional region is defined by the networks and connections between the parts of the region. Another way to conceptualize it is to ask what is the functionality of the people and/or places being interconnected? Geographers often represent functional regions on a map with dots (nodes) and lines, and this leads some to call functional regions nodal. A good example of functional regions is daily newspapers, which commonly have one printing location and a discernable pattern of distribution. Normally, large cities have their own newspaper(s) and it gets delivered away from the city until it becomes too far away for transportation or it competes with an

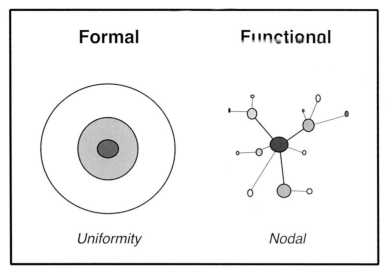

FIGURE 2-3 Region Types

adjacent city's newspaper. Fast food and convenience stores often follow functional patterns based on delivery routes. Think of the fast food chains that only exist in parts of the United States and we see regional patterns. Communication in our modern lives seems boundless, but in reality they are functional regions of connectivity. For example, cell phones require a complex network of actual antennas that bind your mobile conversation in term of the reception quality and availability (number of bars) to nodes of the network. The internet is also a transmission network with actual cables and computer processors, and being "wired" is part of its functionality.

The third type of cultural region is different from the other two because it is a perceptual region based solely on the attitudes and beliefs of the people in them. Vernacular regions are perceived by local inhabitants or members. Generally, they are not labeled on a map or registered as actual place-names. The most often mentioned American example is Dixie. Dixie or Dixieland is not normally found on a map, but many people use the term synonymously with the American South. The names of business establishments that use Dixie in its title or sales slogan reinforce this identification with regionalism. Locally, the use of Aggieland in the College Station, Texas area is very similar. The local inhabitants especially the students of Texas A&M University use the term in a way that demonstrates an understanding of its existence. The "Welcome to Aggieland" sign on the campus water tower and at the Easterwood Airport provide obvious landscape examples. In this case, there's a strong relationship to the student identity label "Aggie" but it also is a simplified way to rhetorically mean the local metropolitan area of Brazos County and the twin cities of Bryan/College Station.

Characteristics of Regions

While regions can be typed into formal, functional, and vernacular, they have some generalized common characteristics. Foremost, all regions are geographical because they exist somewhere on the Earth. The first common characteristic of all regions is they exhibit spatial attributes; spatial is being used instead of geographical to emphasize the measurability of these features. The most frequently mentioned spatial variables are location, area, and borders. Every region has a location; every region has a boundary with the next; every region can be measured. The obvious measurement is area or in the form of a question—how much of the Earth's surface is in this region? We usually answer in units such as acres, square miles, hectares, etc. If a region is properly bounded, one can measure or literally count the things in it such as people and trees. Taken together, we can calculate statistics such as population density, which become useful when comparing regions.

Once we accept the existence of a region, there are two inherent characteristics that follow: internal and external relations. Internal relations are the interactions between the various components of the region. For a region such as Texas, the balancing of actions and interests of its parts can mean multiple layers of dividing. There are 254 counties that effectively subdivide the terrestrial area of the state; there are over 1,200 incorporated places that operate as the local jurisdictions over most Texans. In addition, the internal relations can be between regions as well like East Texas and West Texas. External relations are the interactions between counterpart regions. If we define Texas as a state, Texas has external interactions with other states as jurisdictional equals (not necessarily economic or territorial equals). More than happenstance, Texas closely interacts with the four adjacent U.S. states as direct neighbors. It must interact in significant ways with other large concentrations of Americans (California and New York) for economic reasons. Texas also interacts with Mexico and in particular the four neighboring Mexican states, but this is more complicated because the United States and Mexico (as international counterparts) also have to negotiate the same border.

The final characteristic of all regions is their dynamic quality. Basically, all regions change! Sometimes regions change because they really undergo some sort of movement of people and technology that transform them. For example, the movement of Spanish speakers into the United States is changing the distribution and ratio of English, Spanish, and bilingual speakers. Technology and time has seen the core of cotton production move from East Texas to West Texas. Texas today is different from Texas fifty years ago. Regions evolve due to actual changes, but they also evolve because of the intellectual process. Intellectually, the way we think about regions changes. During the Cold War, geographers saw Kazakhstan as part of the Soviet Union. But after the breakup of the Soviet Union, geographers began to conceptualize Central Asia as a world region. The construction of regions entails an authorial dimension. Region-making requires some generalization of complexity and selection of the more important variables to organize (label) our regions.

LOCATION AND SCALE

Two ideas that are inherent with geography are location and scale. While discussing regions, I've already mentioned location by name and inferred scale while situating region between local and global. It's nearly impossible to discuss ideas in isolation because there is inevitable overlap. Location stands out as a fundamental geographical idea because it is the way we answer the "where-is-it" question. A rather non-technical definition of location is one's place or spot on the Earth.

Location has two connotations. Location can have both an absolute meaning and a relative meaning. Absolute location is exact and precise. Traditionally, geographers have used latitude and longitude to express absolute location. Latitude and longitude is a coordinate system that provides every spot on Earth with a unique address. Currently, UTC and other systems are better suited for the digital age, and they coexist with latitude and longitude. Relative location is not exact and is typically perspectivist. "I live five blocks east of here" is a statement that reflects relative location from some reference point (both the speaker and listener understand). Directions are good examples of relative location as are expressions of proximity or adjacentness. For example, the Rec. Center is just across the railroad tracks; the quickest way to Florida is to take Interstate 10 East. Essential statements about regions are usually relative in nature. Oregon is on the West Coast and Oklahoma is north of Texas are two statements of relative location. Relative location may lack the exactness of absolute, but it works well enough most of the time.

Scale has multiple connotations in geography as well as in general. Consult a dictionary and see all the nuances of definitions for scale. Clearly, the word scale implies a measuring or balancing of weight, distance, and even justice. A generic definition (focused on visual images) is: scale is the relative dimension or ratio of reduction or enlargement in picture, map, etc. In geography, scale is used in two specific ways. The first way is with the use of maps. Cartographic scale, as it is known, is the ratio between the map and the Earth. Every map has a scale so the map reader knows the difference between the map and the real world being mapped. The second way refers to how geographers discuss their subject matter; often geographers claim they can recognize and articulate the nuance of scale. When they do that, they refer to the interplay between the geographical scales of global, regional, and local, which can be either processes of change or resulting landscapes. A geographer might describe Texas blending multiple scales of analysis.

Site and situation is a practical way to exemplify both location and scale. Every place on Earth has an immediate locale (site) and a surrounding circumstance (situation). Site is the local scale, and it includes the immediate terrain, natural resources, and human settlement. Therefore, site is almost synonymous with place, which is the scale that humans interact with one another. Site can also be very exacting in terms of location. Situation, on the other hand, is the relative location of a place. A place's situation can be thought of as one's position to interact and prosper from its connections. Crossroads, ferry crossings, trade centers, and mouths of rivers are generally examples of beneficial situations, but not always because competition and intervening opportunity has to be considered. Furthermore, situations evolve with developing technology and changing transportation patterns, so situation reflects regional and sometimes even global processes.

Let's elaborate on site and situation using some specific and general examples. Galveston was the largest city in Texas for much of the second half of the 19th century. Its site is a barrier island along the Gulf of Mexico, and its situation was the primary gateway for Texas to the world. Galveston's site was ideal for its heyday; it was militarily defendable as an island and economically the best location for transferring between land and water. Unfortunately, the site is susceptible to natural hazards, and it was eventually devastated by a hurricane in 1901. The positive situation of being a migration and trade gateway to the Atlantic world was adopted by nearby Houston. New Orleans, Louisiana also has this basic land-water situation but for a much larger natural hinterland. New Orleans is located very close to the mouth of the Mississippi River as it enters the Gulf of Mexico. Farmers in the Midwest can float their agricultural products downstream very inexpensively, and from there things can be loaded onto seagoing ships. Without a doubt, the situation of being the logical trading and commerce center for the interior United States is superb. On the flipside, New Orleans' site is a precarious one because it is subjected to both fluvial and coastal flooding. The city was originally sited along the banks of the river, the French Quarter, which is the logical high point. Ironically, the flood control structures designed for river flooding contributed to the hurricane-related flooding that devastated the city in 2005.

The relationship between these three, real places is extremely complex and lots of other factors play a role. Some of those factors include happenstance of hazards and boosterism of civic leaders. Think about our university in the hypothetical for a moment. Presumably, every university town has a unique site and story of how it evolved. Universally, there's a surplus of educated people around campuses; some extremely educated professors and a disproportionate number of 18- to 25-year-olds with the best and most recent education. An evolving trend in our current world is "brain power" that includes imaginative ways to construct, tweak, and utilize computers, information, genes, etc. High-tech companies seem to locate themselves in and around the better research universities to take advantage of the super-educated. Instead of natural harbors and cross-roads, research and technology centers give places situational advantages.

MAPS

Maps go with geography. That's about all the rationale necessary to spend some time discussing maps. Geographers use maps in a sophisticated way to illustrate spatial relationships. We've all heard the expression: a picture is worth a thousand words. Maps are worth thousands of words to a geographer.

A well thought out map showing proximity between phenomena, relative location of places, or geography of population density conveys a message that would take pages of writing. Geographers cannot substitute maps for written description, they utilize maps to compliment their writing. Perhaps, the most common experience with maps is with atlases. An atlas is a collection of maps, and these are popular for their aesthetic or encyclopedic value.

Most people will never receive any formal education concerning maps, and in all likelihood, that leads to a little *topophobia* (or fear of maps). On the other hand, most geographers are *topophiliacs* (lovers of maps) and they receive some specialized training. In this section, I will discuss some essential knowledge about maps as well as provide some useful map reading tools. The goal of this section is to bridge the gap between *topo-ignoramus* and *topo-sophisticus*.

The definition of a map is a graphical representation of reality. The three important components to this definition are graphic, representation, and reality. Traditionally, a map was etched into or scribed onto a medium such as paper, which makes it graphical like writing. Currently, many a map are created and used on computers, which would make it a bunch of 0s and 1s in binary code. Nevertheless, we conceptualize maps primarily as visual products. Most people look at a map and imagine something about the Earth. Still, the exclusively visual nature of maps is being rethought—not just by computers. Today, there are maps for the blind that might rely on sound or touch. The second part of this definition is representation; a map is not the thing being portrayed. Like a painting or photograph, a map is both an actual object and a representation of another. Since most maps are showing the Earth's surface, we know they are different objects and different sizes. The final part of this definition is reality. What is real? Reality is but a single reflection of some truth complete with authorial perspective. In the map-maker's mind, she is trying to communicate some fact about the Earth as accurate and precise as possible, but this is limited by her knowledge.

TABLE 2-1 Essential Elements of a Map

	Element	Meaning
1	Title	Topic, Subject or Theme
2	Scale	Ratio / Bar / Verbal Scales
3	Legend	(Key) "Translation of Symbols"
4	Date	Data collection / map production
5	Grid	Index parts of the map
6	Direction	Orientation of map (North Arrow)

There is a special name for those who professionally make or study maps, and they are called cartographers. In fact, some universities actually offer a degree in cartography. In all the instances I know of, cartography programs coincide with geography departments, which reinforces the rationale between geographers and maps. Cartography could be defined literally as mapwriting or more commonly thought of as mapmaking. Yet, a cartographer is more than a mapmaker, they excel in understanding maps as communication devices. Therefore, they not only learn how to design and produce maps, they must evaluate how effectively maps are being understood by their readers.

Map Reading

Let's put the reader into the map reading equation for a moment. Remember, map reading is similar to book reading because it requires a little training and some personal effort to do properly. When looking at a map, there are three different components to see. There are textual, symbolic, and spatial components to a map. The textual component is literally the text that is written on the map. A typical map uses numerous labels for data and place-names. The directive here is to simply read the writing because this requires no special map skills. The second component is symbolic, and this is what often confuses many novices. Maps often have color shades and symbols such as stars, dots, and squares. The cartographer should provide a legend that allows the reader to translate the symbols used on the map. The final component to map reading is the spatial dimension. There are unique and distinct shapes to known geographical units as well

TABLE 2-2 Three Map Types

Map Types	What	Who
General Purpose	Basic Location and information Simplified political and physical Features	Commercial and Government
Topographic	Elevation (contour lines) "earth's surface"	Government Produced Accuracy and Availability
Thematic —isopleth —choropleth —cartogram	Research results "isolines" "areal values" "distortion"	Academics and Others

as the patterns of the specific data on the current map. With spatiality, the task is to recognize the known things and learn the patterns. Most experienced map readers are not conscientiously deconstructing these three components; they look at a map holistically.

Appreciating cartography (mapmaking process and cartographers training) may be useful for the average person to eliminate any apprehension of maps. There are six essential elements to a map that should be present regardless of the topic or theme of the map. Every map should have a title. Basically, this is the big idea of the map. Every map should have a scale, which provides the reader the size difference between the map and the actual Earth. There are different types of scales, so they may not look the same on every map. Every map should have a legend—sometimes called a key. The legend provides a way to translate the symbols used on the map.

Maps should have a date. The date of production is important because the Earth changes and the exact date for when the data were collected is absolutely essential. Most maps will have a grid system for referencing location on the map. Street maps often use a 1,2,3 and A,B,C graph paper-type grid to help people locate a specific street. Maps usually have an indicator of direction, so one can orient the map to the cardinal directions. The most common indicator is a North Arrow or Compass Rose. If there's no indication, by convention, North is at the top of the paper. Both grid and direction can be provided by a coordinate system such as latitude and longitude lines. I will elaborate on latitude and longitude later in this chapter.

Types of Maps

During the course of the semester, I'll use many different maps. I believe it is useful to know something about the different types of maps. There are three basic map types that students should be aware of: general purpose, topographic, and thematic.

General purpose maps are designed to visualize basic location. Most road maps provided by automobile clubs, gas stations, and state governments are general purpose. These general purpose maps are distinct because they definitely emphasize driving landmarks, but they are typical because they simplify both physical and political features onto the same map. An outline map of a country (or even a state like Texas) is typically just a stylistic general purpose type. While features on general purpose maps may be simplified, they can be selectively elaborate and artistically impressive at times.

A second type of map is a topographic map. Even though they show similar features as the general purpose map, they are a special category of maps because they are much more detailed and reliable. The most important factor of a topographic map is that it represents the Earth's shape by documenting elevation. Using contour lines to read the elevation, topographic maps reveal the terrain, aspect, and drainage patterns of the Earth's surface. A close second in terms of importance, is that

36 Part One Introduction to the Geography of Texas

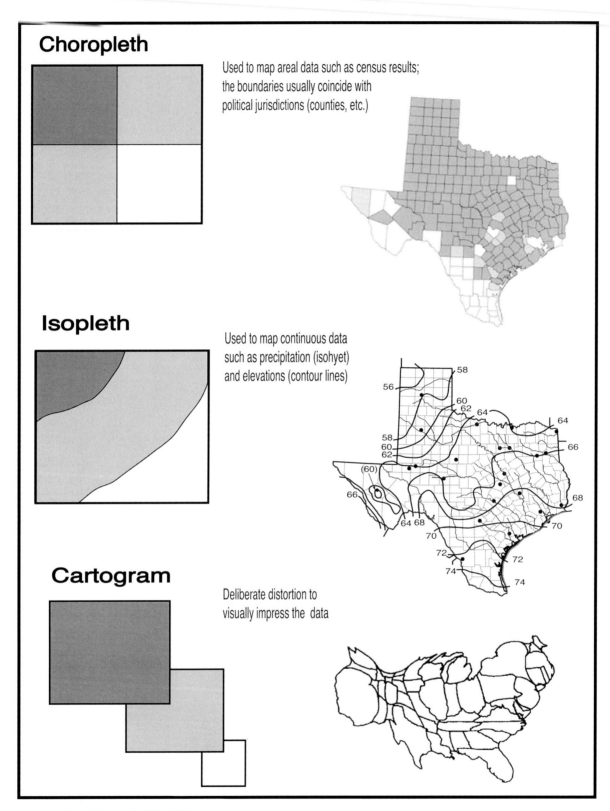

FIGURE 2-4 Thematic Map Types

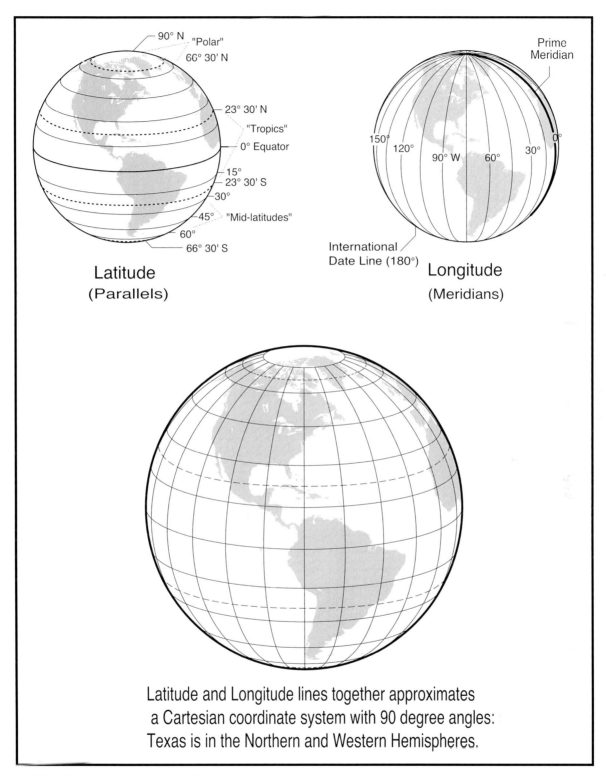

FIGURE 2-5 Latitude and Longitude

TABLE 2-3 Important Latitude and Longitude Lines

North Pole	90° N
Arctic Circle	66° 30' N
Median latitude of Texas	31° 08' N
Tropic of Cancer	23° 30' N
Equator	0° Latitude
Tropic of Capricorn	23° 30' S
Antarctic Circle	66° 30' S
South Pole	90° S
Prime Meridian	0° Longitude
	Naval Observatory in Greenwich, England
Median longitude of Texas	99° 20' W
International Date Line	180° E / 180° W (approximate)

topographic maps are produced by the government, which means comprehensive coverage and high degrees of accuracy. In America, topographic maps are made by the U.S. Geological Service, whereas the Ordinance Survey makes them in Great Britain. Topographic maps have always had a military dimension to them. For example, the types of features such as woodlands, swamps, and built-up urban areas are delineated because they are factors in military movements and operations. Even the colors used in printing the maps were limited so they can be read in low light situations. Another big user of topographic maps is outdoorsmen/women who need the accuracy to not get lost while hiking, hunting, etc. Today, USGS is moving map production toward an all-digital format.

Finally, thematic maps are simply maps that have a theme. Thematic maps are the maps made by geographers to illustrate a research theme. Some examples of thematic maps include precipitation, income, and voting. Thematic maps are topically infinite, and the techniques to make them are numerous but most effective when used with the right data. A quick digression to highlight three mapping techniques is warranted. Three sub-types of thematic maps are isopleth, choropleth, and cartograms. Isopleth literally refers to "lines of equal value," and it is a general term for all isoline maps. Isopleth maps utilize lines to depict some data surface or Z value. Contour lines that depict elevation are technically a type of isoline. Some other examples include precipitation and temperature, which their lines are called isohyets and isotherms respectively. Isopleth maps are preferable for mapping continuous data.

Choropleth maps typically are trying to depict areal values. Most of the time they are using existing boundaries associated with data collection. The census provides a great example because the demographic data is collected by census tract (county, state, etc.), and it is easily mapped using those same geographical units. Choropleth maps are most common with human geography. Cartograms are unique because they purposefully distort geometric shape to convey a message. Typically, a cartogram distorts known units such as states to represent the actual population or some other theme.

Latitude and Longitude

Latitude and longitude are fundamental terms related to various aspects of geography. For instance, absolute location is predicated on the addressing capabilities of latitude and longitude in tandem. Furthermore, orientation and navigation have come to rely on latitude and longitude for referencing location on maps. Two points to make about latitude and longitude are its structure and precision. Latitude and longitude is a Cartesian coordinate system, which is a useful structure. There are other types of coordinate systems, such as polar, which emanate from a single point. Cartesian coordinates refer to a grid-like pattern where vertical and horizontal lines intersect each other at 90° angles. Despite the Earth being spherical, latitude lines intersect longitude lines at 90° angles. Therefore, it is considered a Cartesian system and it can be portrayed on maps as vertical and horizontal lines.

Each place on Earth has a unique address based on two measurements. Bryan, Texas for example, has a latitude of 30° 40' N and a longitude of 96° 22' W, which is pronounced thirty degrees—forty minutes—North and ninety-six degrees—twenty-two minutes—West. Notice both latitude and longitude begin with degrees. Then, each degree can be divided into 60 minutes. Therefore Bryan's latitude is two-thirds of the way between the 30° N and 31° N lines of latitude. Each minute can be further divided into 60 seconds. While not commonly used, a second can be divided, but it is usually articulated as a decimal.

Latitude is a measure of north and south. Technically, latitude is measured as an angle from the Earth's center using the plane of the Equator as the reference. Latitude appears on maps as east-west running lines, which are often called parallels. Lines of latitude are "parallel" to each other because they never cross or converge. The range of possible latitudes is from zero at the Equator to ninety degrees both north and south. Take away the measuring, which is socially constructed, and latitude exists as a function of the Earth–Sun relationship. Low latitudes near the Equator are tropical and high latitudes are polar, which has historically significant connotations beyond their physical properties.

Longitude is a measure of east and west, but there's no natural starting point from which to start measuring. Conceptually, longitude is referenced by north-south running lines called meridians. In contrast to parallels, all meridians converge at both poles, so the distance between meridians change. The measure of east and west must begin with an agreed upon place known as the Prime Meridian. During an international conference, the British standard of using the Greenwich Naval Observatory was adopted. The range of possible longitudes is from zero (Greenwich) to 180 degrees in both east and west directions.

TEXTBOOK DESIGN

In this chapter, we have discussed five fundamental concepts around regional geography. The five ideas or terms to know are geography, region, location, scale, and map. In addition, you as the reader should appreciate that there is a very long intellectual tradition behind these concepts. One studies regional geography for both the practical learning of regional details and for the exposure to the intellectualism of ideas about regions. While you learn more details about Texas, you should grasp that this learning is based on a world of ideas.

The broader purpose of both Chapters One and Two is to introduce students to the geography of Texas. Chapter One hopefully sparked ones attention to Texas and that most basic question—where is it? Chapter Two explored the realm of ideas that geographers operate with when discussing regions. Taken together, Part One is our first steps together to explore Texas.

The final component to Part One and this chapter is to outline the rest of the book. I initially organized Parts Two, Three, and Four along Carl Sauer's approach to cultural landscapes: human agency (culture) and natural landscapes (medium) mix to produce the cultural landscape. Broadly speaking, Part Two will be about the culture-history of Texas highlighting the cultural diversity of the state. The human agency of Texas' past is explored with the geographical ideas of diffusion and migration. Part Three will focus on the physical geography of Texas, which has extremely diverse natural landscapes. While thinking of the physical environment as a medium, the outward forms are emphasized over the understanding of physical processes. Part Four returns to the human geography of Texas and some of the actual landscapes we can see today. A few of the highlighted themes are organization and structure that contribute to our contemporary landscapes as well as mediate our experiences in them. As a way of concluding, in Part Five, we return to the major regions of Texas by approaching them in a regional fashion. In addition to concluding our exploration its we can contemplate future geographies of Texas. Where will Texas be in the near future? Only time will tell. Next, we turn to the past.

"The Road That Started It All"

FIGURE 3-1 Texas 21 Sign in the Pineywoods
El Camino Real de los Tejas, Cherokee County. Photo by Author.

Part Two
Cultural-Historical Geographies of Texas

INTRODUCTION TO CULTURAL-HISTORICAL GEOGRAPHY

Part Two of *Geography of Texas* is a look into the human past. Primarily, Texas is a human region because of the people who migrated here and transformed the social relations and natural environments into what we see today. While the three chapters have distinct topics, the broadest unifying theme of this part is that culture modifies the Earth. In particular, the focus is on historical culture groups—each with their unique human agency or ability to make change. Humans move about on the Earth's surface and transform both the Earth and themselves as they live their lives. This transformation of the Earth is observable in the cultural landscapes that previous people left behind as well as the landscapes current inhabitants live in.

The motivation for altering the Earth is both basic survival and well-being. In a literal sense, we make hearth, build shelter, or construct home. Additionally, we hunt, gather, and store food, water, medicine, etc. Beyond basic shelter and nutrition, we do much more in the form of communicating and interacting with each other. The uniqueness of being human is our language. We develop and share our collective wisdom through language, which might be our most important tool. We also use language to distinguish ourselves, construct our identities, and create places. Therefore, geography sees both a universal dimension of culture at the global scale and a local uniqueness that distinguishes groups of people and their places.

The story of Texas as a region is some what complex a sit is interconnected to events and processes at different regional scales. In particular, the early success of Spanish colonization. The 'European rivalries, and eventually the interwoven pathways with an expansionist UnitedStates.

Part Two begins with a chapter on the historical geography of Texas that encapsulates all of the past up to statehood. The second chapter looks at the cultural geography of the state, which has been a characteristically diverse realm. Finally, a chapter on the cultural landscapes of Texas explores some of the symbolic and iconic landscapes that one finds around the state.

HUMAN MOBILITY

In the big scheme of things, humans are very mobile beings. We may not be the fastest terrestrial species nor are we likely to travel extremely far in a single journey. We definitely move around. In fact, we are biologically designed to be mobile. We stand erect on our two feet, which is called bipedalism, and we propel ourselves over the Earth's surface. With this ability to walk and run and even swim, humans explored beyond their known worlds of immediate surroundings and past experiences. We humans have enhanced the movement by riding animals, pedaling bikes, and eventually driving automobiles. Ships, trains, and airplanes have given modern people even more potential mobility as well as the more expansive ability to quickly exchange goods and services. In short, we like to move around, and it could be argued that mobility has become part of human culture. One only has to think about historic Route 66 to see how popular culture and mobility go together in this articular American example. Geographers have two important concepts to help understand mobility: diffusion and migration.

Diffusion is the geographical spread of ideas, innovations, and attitudes through human contact and communication. One of the assumptions is that new ideas and inventions only occur once in one place. Then, other people and places are exposed to the new thing, and they have to decide to adopt or reject it. While cultural diffusion is occurring, its expansion follows a pathway from the point of origin to its areal extent that can be dated and mapped. Given enough time, the idea, innovation, and attitude can diffuse around the entire world. Some classic examples of diffusion include the invention of the plow, domestication of animals, and the emergence of writing. Popular culture also has its examples such as blue jeans, hula hoops, and i-pods.

Academics can model diffusion as a way of better understanding this geographical phenomenon. *Expansion diffusion* is the spread of ideas, innovation, and attitudes as the actors involved are stationary on the Earth's surface. Most of the time, we see expansion diffusion as neighbors or those closest to the new thing simply being exposed and adopting first; therefore, proximity matters. In some instances, diffusion follows a pattern associated with people being connected or having privileged access to the new thing. Imagine a major news event as the breaking details are being broadcast on TV, those watching are exposed first and simultaneously with little regard to actual proximity. *Relocation diffusion* occurs when people move on the Earth's surface. Not only do people carry their material belongings, they bring their experiences and knowledge in their minds as they move. As migrants move to another place, they diffuse their ideas, innovations, and attitudes with them. In many instances, diffusion occurs with a combination of both expansion and relocation.

Migration is the relatively permanent movement from one place to another. In contrast to short-term movements and temporary situations, migration entails an implied permanency. In the world today, there are numerous people who are displaced or in a refuge status, but their movement isn't considered migration unless they permanently settle away from their homeland. Scholars prefer to think of migration as the product of a voluntary decision, and the ideal situation is an informed decision between two places. Migration under these ideal conditions is also known as *voluntary migration. Involuntary migration* is another current and historical reality. The involuntary nature means forced or coerced in some manner. Modern refugee situations often include this dimension where people have to flee because of war or famine. In the context of America, the movement of African slaves is a larger example that changed the human geography of the Western Hemisphere.

Place-utility theory says that people will make rationale choices to migrate. Push factors are things that make people leave a place; for example, poor socioeconomic conditions and political oppression are factors that make place A undesirable. Pull factors are things like better prospects or opportunities that draw people toward place B. It is entirely possible that one's impression of the other place (pull factors) may not be accurate and based on misinformation. In place-utility theory, when the pull factors of place B balance favorably with the push factors of place A, people make the decision to move. It is entirely possible that in the process of migrating, an intervening opportunity (place C) may occur and the migration ends up there instead. Migration is a big part of contemporary culture in places like Texas because there is a steady influx of migrants that choose to live here.

LEARNING OBJECTIVES: CULTURAL-HISTORICAL GEOGRAPHY

Part Two of this textbook is associated with our past: the historical geography and cultural geography of Texas.

- Discuss the geographical aspects of Texas history

    You should be able to identify significant places from Texas history and understand their locations.

- Discuss the cultural diversity of Texas

    You should be able to identify and elaborate on the major culture groups in Texas and know their relative locations.

- Discuss the cultural landscapes of Texas

    You should be able to elaborate in some detail on the exemplified landscapes: ranching, cemeteries, etc.

- *Maps:* Familiarity with anthropocentric maps of Texas

    You should become comfortable using historical maps that portray Texas with different perspectives and borders:
    - Spanish Texas
    - Mexican Texas
    - Republic of Texas

    You should be able to read cultural information from relevant maps.

- *Definitions:*
    Culture
    Landscape
    Diffusion
    Migration
    Language
    Religion
    Settlement
    Sequence occupancy
    Cultural Hearth

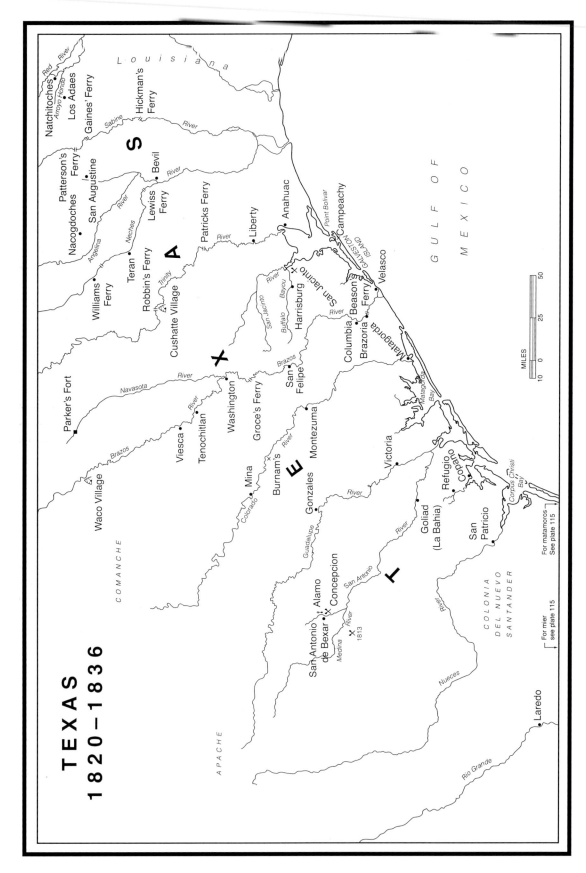

FIGURE 3-2 Mexican Texas: Castenada's map of Texas, 1820–1836
Carlos E. Castañeda's depiction of Texas during the Mexican Era. Previously published in the *Atlas of American History*.

3
A Geographical Past

Texas did not exist five hundred years ago, yet there were people. There just was no conception of political boundaries, organized trade, and professional rodeos that we could call Texas. Half a millennium ago the natural environment was not all that different than it is today in the big scope of geological time. The rivers, mountains, coastlines, forests, and grasslands were all in their current places. Indigenous people adapted to all the different environments found in Texas, and simultaneously, they modified the environment through their activities. The population density was not high by global standards, so the modifications were not dramatic when compared to other regions of the Americas. As new culture groups with higher technological abilities and increasing populations displaced Native Americans, the environmental changes become extremely profound.

Chapter Three examines the region as it became Texas; it documents the human history and the political evolution up to statehood. First, we recognize the long human presence in Texas; second, we acknowledge the fundamental transformation of the Americas that came with European contact; then, we outline the successive political regimes including the Republic of Texas that still contributes to the contemporary cultural identity.

CLOVIS

There are big debates about when humans arrived in the Americas. Most scholars think that humans migrated from Siberia (Northeastern Eurasia) to Alaska (Northwestern North America) during the later stages of the Pleistocene. As the Earth was coming out of its most recent Ice Age, sea levels were much lower, which would have allowed a "land bridge" or non–sea-based migration route between the continents. Then these modern humans migrated southward and eventually occupied all of the inhabitable parts of the Americas. An obvious precondition for this scenario is a human occupation of Siberia with the cultural development to survive and even thrive in the sub-polar latitudes and periglacial climatic conditions. Plants and animals of the high latitudes have evolved together and contiguously around the circum-Arctic, which humans could learn to utilize in just one part. Most proponents of the land bridge thesis assume humans followed their hunting game into the Americas, and they could explore and expand into an already familiar environment (pre-adaptation). A second issue is the speed of migration throughout the Americas, which appears to be rather quick considering the extreme range of north to south environments. In actuality it averages only about 10 miles a year if we calculate the distance between the Alaskan Peninsula and Tierra del Fuego and divide it by a millennium. It is reasonable to think that modern humans would have the ability to expand at that rate even if we don't know the exact routes or specific moments.

Contrary interpretations are common. Recent speculation includes that this early intercontinental migration could be done by sea travel in small distances along the Bering Sea Coast. This scenario allows for a slightly different timing that doesn't rely on the land bridge being present, but fundamentally follows the same pattern of a north to south occupation. It may be possible that the migration could be by both routes simultaneously, which allows for a more complex culture with both land and sea-based mobility and resource use. Humans occupied New Guinea and Australia by boat much earlier, perhaps 40,000 years ago (or BP—before present); therefore, the ability to do so cannot be discounted without careful consideration. This fact seems to inspire alternative theories of earlier human presence in the Americas, which could have occurred by purposeful or accidental sea travel. Theoretically, humans could also have crossed the Bering land bridge during an earlier Ice Age or low sea level event.

The great controversy in American archaeology is the dating of earliest humans in the Americas. Every couple of years or so, somebody announces a discovery that would upset the generally accepted dates and/or story line of a southerly migration in the 12,000–15,000 year range. After closer scrutiny, most of these sites have methodological problems that other scientists can reasonably interpret different results. Taken in the entirety, some evidence exists for a possible earlier human presence around 40,000 BP despite no single site with incontrovertible evidence. These sites and theories do not prove a direct ancestral link with Native Americans nor even any continuity of cultural artifacts and traits. Furthermore, the lack of widespread archaeological evidence for the time frame is disturbing. In Europe, Cro-Magnon and Neanderthal sites for the same time frame are numerous and widespread, so they should be available in the Americas if humans did occupy the area.

One of the significant archaeological sites is Clovis. While Clovis is not the oldest undisputed site, it is talked about as a conventional measure of time. The convention is to name sites according to their location and if an initial discovery is made there, the site name becomes synonymous with the discovery. At Clovis, which is on the New Mexico side of the Llano Estacado, scientists unearthed arrowheads that could be accurately dated to 13,000 BP (11,000 B.C.). These arrowheads are usually called Clovis points. Additionally, "Clovis" is used as a surrogate for time in the archaeological debates. For example, the discovery of a 20,000 BP site would be called a possible "Pre-Clovis"; whereas a 10,000 BP site would most likely conform to our Clovis perspective.

Regardless of the debate, Clovis puts humans in the area that would become Texas at approximately 13,000 years ago. The resulting conclusion is that humans have occupied and interacted with the environment for a long time compared to the relatively recent, contemporary period of European dominion and American sovereignty. During this long period of time, the initial inhabitants modified the environment, and various cultures can be identified. For example, it is becoming clear that indigenous people used fire throughout Texas to improve grassland conditions for their hunting prey.

COLUMBIAN EXCHANGE

The human occupation of the Americas and the subsequent parallel developments that occurred set the stage for a most dramatic event. Alfred Crosby coined the term Columbian Exchange to reflect the totality of the interchange of the last 500 years. Previously, Columbus' voyage of 1492 was commonly called the discovery of the Americas. While contemporary scholars use alternative words such as encounter, the significance of this exploratory action is still enormous. Prior to his voyage, most people were ignorant of the Earth's spherical shape, the composition of continents, and the collection of world cultures. Afterward, educated people could conceive the whole world as it actually exists; Europeans could revel in the discovery of new people and places and the opportunities to interact with it. Therefore, 1492 becomes an exact date to measure the starting point for contact and communication between the Americas with Europe, Asia, and Africa, and Columbus becomes the central, historical figure of the discovery. Furthermore, this contact/exchange coincides with the economic, ecological, and technological developments in Europe that would propel it to an advantageous relationship with the rest of the world, which is identical to the story we tell when trying to understand globalization.

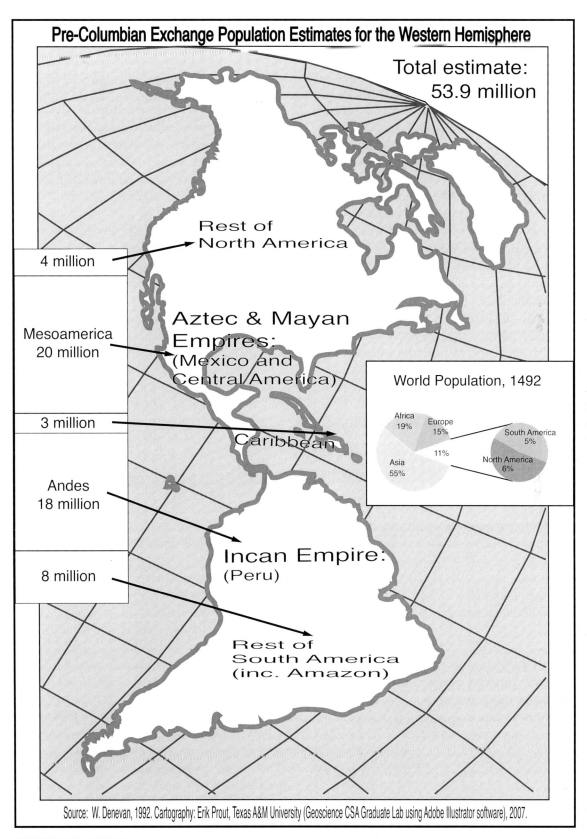

FIGURE 3-3 1492 Americas

The definition of Columbian Exchange is the two-way exchange of people, plants/animals, and cultures between the New World and Old World since 1492. Although Europeans may have been initiators and shapers of this exchange, it is a two-way process. People from both "worlds" migrated to the other, and in some instances large-scale mixing occurs. Mexico is one place where this mixing is part of its national identity, and Mestizos are the statistical majority.

Domesticated plants and animals were deliberately transported and eventually adopted by culture groups that were unaware of these prior to 1492. For example, corn, tomatoes, and chili peppers were initially domesticated in the Americas, but now can be found around the world. In fact, it's hard to imagine Italian cuisine without the tomato or Thai cuisine without the chili. These examples show how complete the exchange is when we collectively forget the origins of basic food products. Likewise, Old World plants and animals diffused to the Americas. Domesticated species such as cows, pigs, sheep, and horses were introduced by Europeans as well as wheat, rice, and apples. In addition, ornamentals and pets were brought in purposefully; unintentional introductions such as earthworms were also common. At a microbiological level, a whole new exchange of diseases and germs occurs, which will have unequal consequences. Finally, a wholesale exchange of cultures diffuses across the oceans. Another way to view culture is knowledge or know-hows. The knowledge of how to use resources, make tools, and communicate with one another. The diffusion of cultural features such as language and religion is strongly asymmetrical as European languages and Christianity are implanted in the New World. Nevertheless, indigenous American knowledge diffuses to Old World migrants firsthand and eventually diffuses with adaptation, corn for instance, in the rest of the world.

Europe, Asia, and Africa (and technically Australia) comprise the Old World where contact and communication were consistent and of long duration. The "Out-of-Africa" thesis for human origins and occupation of the planet states that humans have migrated from Africa in successive waves to Asia and Europe. The larger point to make is that there are no major barriers for humans, plants, and animals, therefore biogeographical patterns reveal this interconnectivity. There are still differences between things, for example, the African elephant and Asian elephant, but there is no equivalent in the Americas. As cultural innovations such as agriculture and writing are invented, they in all likelihood diffuse to the other continents given enough time.

Except the Americas! North and South America evolved from a different set of plants and animals, so Native Americans had a different tableau for domestication. Agriculture as permanent settlements and investment of time in growing crops and tending animals is a relatively recent human development of the last 10,000 to 12,000 years, which is just after human occupation of the Americas. Modern humans in both the New and Old Worlds developed agriculture independent of one another. Both worlds had cultural hearths and sophisticated civilizations where many agricultural innovations initiated. Yet they differed in one significant aspect, they evolved different disease complexes. Old World domesticated animals, for example, cattle, pigs, and chickens, interacted with their human handlers and shared diseases that were much more deadly than New World counterparts. Think back to your own childhood immunizations, we still inoculate for measles and small pox. While Old World cultures had developed enough resistance to these deadly diseases to make cohabitation with these animals a positive relationship, New World people had never been exposed to them.

The immediate response to contact and interaction was a devastating reaction to disease. While not being understood in today's scientific knowledge, it was apparent that something was going horribly wrong for Native Americans. As Europeans explored and settled in the Americas, they were inadvertently introducing deadly diseases to the indigenous populations. The spread of disease and the subsequent high mortality rates did more to affect the geopolitical situation than any other single factor including technology like guns. Indigenous groups were repeatedly exposed to different deadly germs that resulted in an overall mortality rate of close to 90 percent. Besides the obvious demographic collapse, many of these culture groups experienced a profound social collapse as leadership and knowledge was lost. Their ecological imprint was drastically reduced; the ability to resist cultural change was diminished. For all practical purposes, indigenous America was devastated. Europeans misinterpreted this decline as an unequal humanness, which corresponds to the racism that they exhibited toward Africans and Asians. A huge consequence of this population decline is that they have a virtual *"tableau blanco"* or blank slate for imposing their cultural norms onto the New World. Quite literally, Europeans could migrate to the Americas, introduce their landscape preferences, and create whatever they wanted because there was no serious resistance, and they believed that they were entitled to do such. In addition, they wanted a new labor source to replace the indigenous people, and the slave trade was initiated. In summary, the Columbian Exchange was a monumental event in world history that had enormous demographic, cultural, and ecological consequences. Europeans managed to ride this change on top; however, Native Americans struggled to survive and adapt.

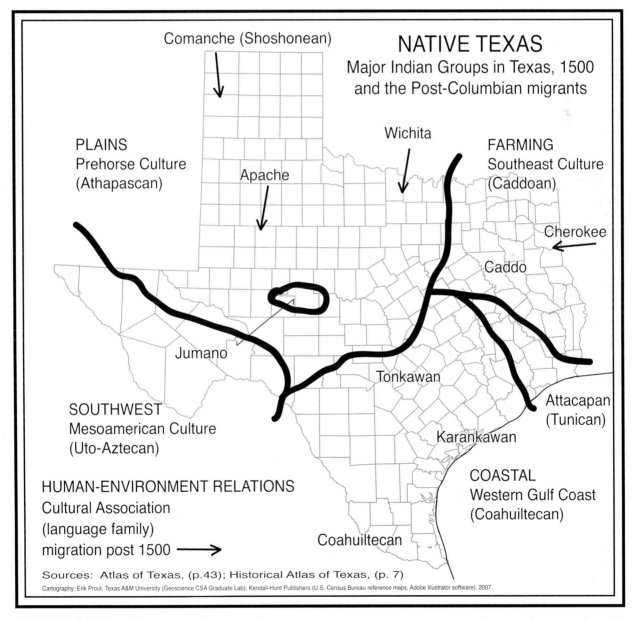

FIGURE 3-4 Native Texas

NATIVE TEXAS

Native Texas is a contemporary idea to describe the indigenous past in the region that we now call Texas. While scholars assume the presence of humans goes back to the earliest occupation, the evidence is spotty and the argument relies on reasonable interpretation. Furthermore, the changing spatial patterns of cultures as they replace each other over this long period of time are vague. The indigenous patterns at the time of contact are better known, and we can classify them accordingly. Broadly speaking, three human-environment communities existed at contact: farming, coastal, and plains.

The farming communities were the more developed societies because, as implied in the title, they were agriculturalists. These complex farming communities were located in the eastern side of the state. As farmers, they modified the Earth's surface for optimal harvests, and they settled near by to watch/protect their investment of time. Therefore, they were sedentary groups that relied on their ability to grow crops, and they constructed relatively permanent dwellings and ceremonial sites. The culture groups we associate with East Texas are the Caddo. It turns out the Caddo are more of a confederacy of closely related tribes with friendly relations to one another. It is the Caddo word for ally (or friend), *Tejas,* that eventually becomes the name for the entire region.

The Caddo and sedentary farming cover only a small part of Texas; the other two types covered more surface area. Various coastal tribes occupied much of the coastal plain stretching along the Gulf Coast from Galveston Bay to Corpus Christi Bay and reaching inland to Hill Country. These communities tended to be semi-sedentary staking out a river valley as it approached the Gulf. These tribes could take advantageous of two distinct resource bases: coastal fishing and shellfish collecting, and hunting/gathering along the fresh water rivers. The Coahuiltecan were located along the Rio Grande; Tonkawa were more inland; Karankawas were along the coastal bend. The remaining community type is the Plains Indians.

FIGURE 3-5 Caddo Mounds in East Texas
Caddoan Mounds State Historic Site, Cherokee County. Photo by Author.

These tribes were migratory and generally moved southward on the Great Plains. They hunted buffalo and other big game and probably migrated with the cycles of their food source. The tribal names scholars associate with Texas are the Apache who dominated the Southern High Plains and the Jumano in the Southwest with overlapping presense in Northern Mexico

Before contact, Plains Indians did not have horses, so our stereotypical image of the Plains Indians is a post-contact reality. Native American culture groups migrated and overtook each other prior to contact, and in the last 500 years we can identify this process. Two examples in Texas that occur after contact but not necessarily because of it: Comanche and Wichita. The Wichita expanded southward into North Texas. Some scholars consider the Wichita to be related to the Caddo; sometimes even called the Northern Caddo. The Comanche migrated southward into the Panhandle and West Texas displacing the Apache. It is the Comanche that adapted horses and rifles and eventually dominated the southern Great Plains.

Scholars associate other Native Americans with Texas as part of a general displacement of Indians from the Southeastern United States. A sequence of groups mostly on their way to the Federal Indian Territory (Oklahoma) passed through Texas and a few eventually stayed. The Cherokee were one of the Five Civilized Tribes that quickly adopted European ways, nevertheless, the American government forced them off their lands. Some of the Cherokee ended up in Northeast Texas and stayed for a few generations, but they eventually moved to Oklahoma. Smaller groups such as the Alabama-Coushata and the Kickapoo were granted small reservations and remained here. On the other side of the state, the Tigua currently have a small reservation in the El Paso area. The Tigua moved down the Rio Grande from New Mexico with Spanish missionaries during the Pueblo Indian Revolt, which would become known as the oldest European settled place in Texas.

In conclusion, native Texas has four important geographic points. Firstly, there was contact and communication between Europeans and Native Americans. It is possible that indigenous knowledge such as crop types and tillage techniques was given to early missionaries. The enduring legacy is that native tribal names were adopted as place-names that are still in use today—including most notably Tejas/Texas. Secondly, Texas was adjacent to the Federal Indian Territory, which later became the state of Oklahoma. Thirdly, Texas was also a migration destination, primarily with tribes that were being relocated from the American Southeast. Finally, the amount of land devoted to Indian reservations is very small when compared to other states. The best explanation revolves around land ownership. Reservations in the United States were a function of federal government policy, and unoccupied land in Texas was controlled by the republic and state governments who preferred to sell land to finance government.

MESOAMERICA

Mesoamerica was a series of civilizations that occupied the sub-tropical landmass of North America, which corresponds to modern-day Mexico and Central America. Cultural geographers count Mesoamerica as one of the important global cultural hearths where innovations and ideas originate. Many scholars think Mesoamerica had one of the better diets in the world before the Columbian Exchange with its trilogy of corn, beans, and squash. Clearly it supported large populations, perhaps 20 million in 1492, and the cultures created numerous symbolic landscapes such as pyramids. The Mayan Empire preceded the Aztecs and was located on the Yucatan lowlands as well as in the mountainous highlands. The Tolmecs were centered in the Valley of Mexico. It was the Aztec Empire centered in Mexico City that the Spanish encountered and Cortoz conquered in 1521. After contact, Mesoamerica remained populated with significant concentrations of indigenous groups.

SPANISH TEXAS

While Christopher Columbus was from Genoa in Italy, he explored for the Spanish. It is the Spanish who first ascertained the presence and significance of the Americas. They used their head start to determine where the existing civilizations and wealth were located, and they concentrated their efforts on those places: specifically Mexico and Peru. Mexico was understood to be the biggest prize for the Spanish who honorifically called it New Spain. The area we now know as Texas fell into the New Spain sphere of interest. Drawing on its experience from the *Reconquest* of the Iberian Peninsula, Spain assertively imposed itself on the Americas with its ideas and authority. This Spanish imprint was most noticeable in those places they concentrated their efforts; eventually a standardized plan for shaping the land and the people diffused throughout what we now know as Latin America. Some examples of this imprint are the central plaza and grid pattern of streets which produce the most common city pattern in the Americas, and typically the largest Catholic Church sits adjacent to the square-shaped plaza.

The earliest European contact with the area we now know as Texas was by Spanish explorers in the first half of the 16th century. Pineda was the first European to see the Texas Coast when he sailed around the Gulf of Mexico in 1519. In the next couple of decades, overland expeditions crossed over Texas, and they recorded information about the frontier. During this exploratory phase the Spanish claimed the vast areas north of Mexico, but no permanent settlement was established. The first regular contact and eventual settlement came not from the coast but inland from interior Mexico. As miners and missionaries slowly expanded northward, they traveled by way of the upper Rio Grande to the Pueblo Indians in current day New Mexico. When the Pueblo revolted, Spanish officials and some loyal tribes retreated to El Paso and they established a mission, *Ysleta,* which became the first permanent European settlement in Texas.

Spanish interest along the coastal plain was minimal, and they were reasonably content with a vague frontier until another European power appeared on the horizon. While the French never seriously claimed Texas, their unplanned, failed attempt to create a settlement along the coast got Spanish attention. The historical figure that scholars focus on is LaSalle because he explored and claimed for France the entire Mississippi River System, named for the king—Louis. On his return voyage to create a settlement (New Orleans) in 1685, he missed the mouth of the river and eventually landed in Matagorda Bay. They decided to stay put and founded Fort St. Louis, but it failed. Spanish officials were alarmed by reports of the French, and they immediately set out to destroy it in 1689. Since the settlement had already failed, they obliterated the evidence so as to remove any possible French claim. The strategic plan was to set up a forward frontier presence—lacking any established border—as far east as possible. The normal Spanish colonization scheme includes a military presence (presidios) and a governance/urban dimension (cities). In this instance, the primary effort was with missionaries, who tried to implant Spanish influence through contact with the Caddo Indians. Eventually a forward capital was established at Los Adaes; it reality in was only a small outpost along Arroyo Hondo close to the French trading outpost at Natchitoches on the Red River. A long overland trail was formally established with Saltillo, which necessitated way stations and river crossing points. Approximately halfway between them, San Antonio developed from a way station into a coherent Spanish settlement with missions, a presidio, and eventually a city charter in 1731. The road that linked Saltillo to Los Adaes became known as the San Antonio Road; it is often referred to El Camino Real because it was considered an official road and now parts of it are signed with OSR.

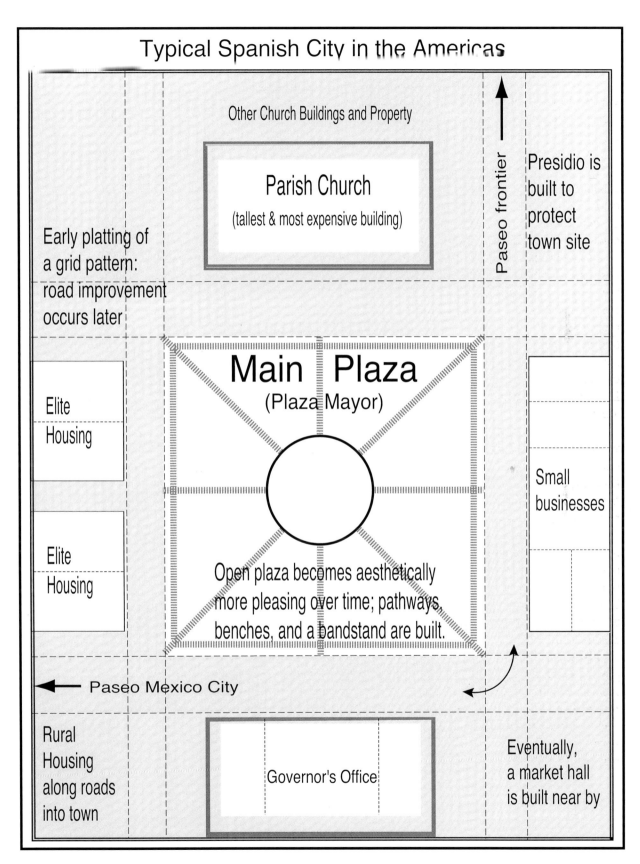

FIGURE 3-6 Spanish City Model

FIGURE 3-7 Spanish Plaza, San Elizario
San Elizario Mission, El Paso County. Photo by Author.

SPANISH LOUISIANA

Texas' eastern boundary was never defined formally between the Spanish and French. When the French surrendered to the British in 1763 (French and Indian War), they renounced their claims to North America. French Canada was given to the British, but they transferred Louisiana to the Spanish. Because the whole purpose of having a frontier presence was removed, Spanish moved Texas provincial government to San Antonio. Spanish officials occupied New Orleans and reassessed their trade patterns. They opened up Texas to New Orleans that permitted the movement of people and trade to go East instead of overland to Northern Mexico. The earliest cattle drives out of Texas initiated here along the Opelousas Trail with New Orleans as the destination.

During the Napoleonic Wars, Spain returned Louisiana to the French under duress. Napoleon sold Louisiana to the Americans a few years later. The result was that New Spain and the United States became neighbors without a clearly defined border. As military conflict appeared certain, commanders in the field created the Neutral Ground agreement that kept U.S. troops East of the Arroyo Hondo and Spanish troops West of the Sabine River. In 1819, the governments formally signed a treaty (Adam–Onis) that defined an exact border. That border would eventually define four of Texas' nine border segments.

Spain's dominion over Texas lasted for 300 years; Spanish efforts to populate the area were rather feeble. Overall, Spanish settlement of Texas was a failure. After 300 years, the nonindigenous population was only around 4,000. Spain could never garner enough interest or persuasion to get many Spanish citizens to migrate and settle in Texas. They hurt their own efforts by constraining trade and communication through central Mexico, albeit this started to change when they controlled New Orleans. Nevertheless, Spain did lay out the framework for future development by site selecting roads and towns on the coastal plain.

Three exceptions to settlement failure were found on the coastal plain: San Antonio, La Bahia, and Nacogdoches. San Antonio was the emerging main place for this Texas province. La Bahia was a cluster of missions downstream from San Antonio that will form into a settlement known as Goliad. Nacogdoches was established from the previous Los Adeas settlers who had petitioned to return to the Piney Woods. Although continuous settlement was occurring in the El Paso area, Spain considered that to be part of Chihuahua and their interests in the upper Rio Grande. "Texas" was the buffer zone between northern Mexico and French/American Louisiana and their notion was confined exclusively to the coastal plain.

In addition, it is clear that Spanish culture diffused to Texas during this early European dominion. Two examples of Spanish culture are ranching and architecture which reveal landscape continuities with the northern Mexican states and territories. Significantly, the Spanish released Old World animals such as cattle, sheep, pigs, goats, and horses into Texas for the purpose of ranching and eventually setting. The Spanish vaquero preceded the Anglo cowboy by a long shot. Building techniques as well as city design are introduced by the Spanish. Plazas and grid patterned settlements are seen in Texas prior to 1821. Yet one must be cautious because we know that Spanish culture is being reinvigorated with more recent Mexican migration and diffusion.

MEXICAN TEXAS

Mexican Independence means that all the people and territory of New Spain are now locally governed by an "American" entity and not a European monarchy. The initial declaration of independence was issued in 1810 by Mexican intellectuals, and the formal recognition of independence by Spain occurs in 1821. In the early years there was a lot of political jostling about how to define Mexico and how it should be governed. The Constitution of 1824 created a federal system very similar to the United States actually called the United Mexican States composed of 19 states and four territories. Instead of calling Texas a territory with a pathway to statehood, the constitution conjoined Texas to the state of Coahuila. Three intertwining aspects of Mexico's 15 years of sovereignty over Texas stand out: Empresarios, settlement/population changes, and governance.

Empresarios began the process of large-scale land titles and organized migration to Texas. Empresarios are a Spanish system for granting an individual or corporation a charter to bring in settlers. The person receiving the charter is called the Empresario, who typically acts like a real estate agent, and the area being settled is usually named after them. The Empresario is responsible for successfully settling a set number of people/families and in exchange they receive a large amount of land for their effort. The best known Empresario was Stephen F. Austin although in English, it is usually translated as the Austin Colony. Actually the genesis of the Austin Empresario dates back to his father, Moses Austin, who was negotiating with Spanish officials at the time of Mexican Independence. Upon his death, Stephen successfully finalized the deal in 1821 and he began to recruit suitable families to settle an area between the lower Brazos and Colorado Rivers. The original settlers became known as the Old 300, in fact, Austin was one of the only Empresarios to properly fulfill his charter.

The Empresario system expanded to other agents and geographical areas of the state. Both the De Witt, centered on Gonzales/Guadalupe River, and De Leon, centered on Victoria/Lavaca River, formed in 1824; Irish Empresarios in San Patricio and Refugio formed along the Nueces River. Much more confusion surrounds the East Texas Empresarios because unofficial agents were selling false titles, American/French squatters were de facto settlers, and the Mexican authority was minimal; eventually, the Galveston Bay and Texas Land Company consolidated these holdings. In addition to initiating the idea of private property and diffusing Spanish land and water rights to Texas, Empresarios produce one additional legacy. The Empresario system translates into large blocks of land that are settled with a master plan imbedded with a Cartesian order and structure. Therefore, cartographically designed cities, such as San Felipe, Gonzales, and Victoria, appear. One sees large-scale property surveying, which extends into regional political boundaries and there are differentiations of land use. Lastly, these master planned communities reveal an emerging ethno-cultural segregation when Empresarios controlled who migrated.

As migration to Texas increased, the settled parts of the state expanded. In addition, the composition of the settlers evolved toward an Anglo-American majority. For the Mexican government, the unsettled frontier between Central Mexico and the expansive United States was of great concern. The Mexican government recognized the Adams–Onis Treaty as their boundary, but they knew that the relatively uninhabited North was at risk because they couldn't militarily prevent attack. They took a big risk allowing settlers from the United States, and they hoped the Empresarios would control the migration by selecting only good families who would remain loyal to the Mexican government. These efforts to populate the frontier were reasonably successful, but the percentage of new settlers was disproportionate toward the American South who brought slavery and their politics. The settled area of Texas was focused on the coastal plain,

FIGURE 3-8 Austin Monument in San Felipe
Stephen F. Austin State Historic Site, Austin County. Photo courtesy of Texas Department of Transportation.

and much of the settled areas were on coastal side of the San Antonio Road. The threat of Indians prevents any substantial settlement on the inland side of the road. After the success of the Austin and De Witt Empresarios, further development moved toward the coast. Small places that serve travelers at ferry crossing sites appear as well as the emergence of ports along the coast. In addition, Americans are expanding from Louisiana and Arkansas in an unorganized fashion, and along the Red River, some are inadvertently crossing over into Texas to settle.

By 1830, the ratio was nearly 4:1 Anglo-American, and Mexican authorities began to be concerned. Some of their responses resemble current American reactions to Mexican migration. In the 1830s, Mexico tried to halt uncontrolled migration from the United States while they tried to regain control of the process. This actually backfired as illegal movement continued, which happened to coincide with the previous Neutral Ground Zone, and a lawless environment took hold in parts of East Texas. The biggest grievance was that Texas was conjoined to Coahuila, which from Mexico City was a logical decision to attach a potentially restless frontier with an established Mexican state. Distance was still a significant factor in travel and communication, and that means for a citizen in San Felipe the place to file records in Saltillo was over 500 miles away. Nevertheless, there was actually a lot of governance evolution during the Mexican control over Texas. First, they introduced municipalities to Texas. The initial four places were Bexar, La Bahia, San Felipe, and Nacogdoches. By 1834, there were 13 recognized municipalities, and 10 more were formally formed in 1836.

Municipalities were actually more like central places—hinterlands that combine a city and its surrounding countryside into one administrative unit with commissioners *(Ayuntamiento),* chief executives *(Alcade),* and a legislative council *(Regidores).* When Texas became independent, municipalities became counties, and some of the original structure survives. Acknowledging Anglo concerns, Texas was subdivided into three departments in 1831 to better serve local constituencies without providing

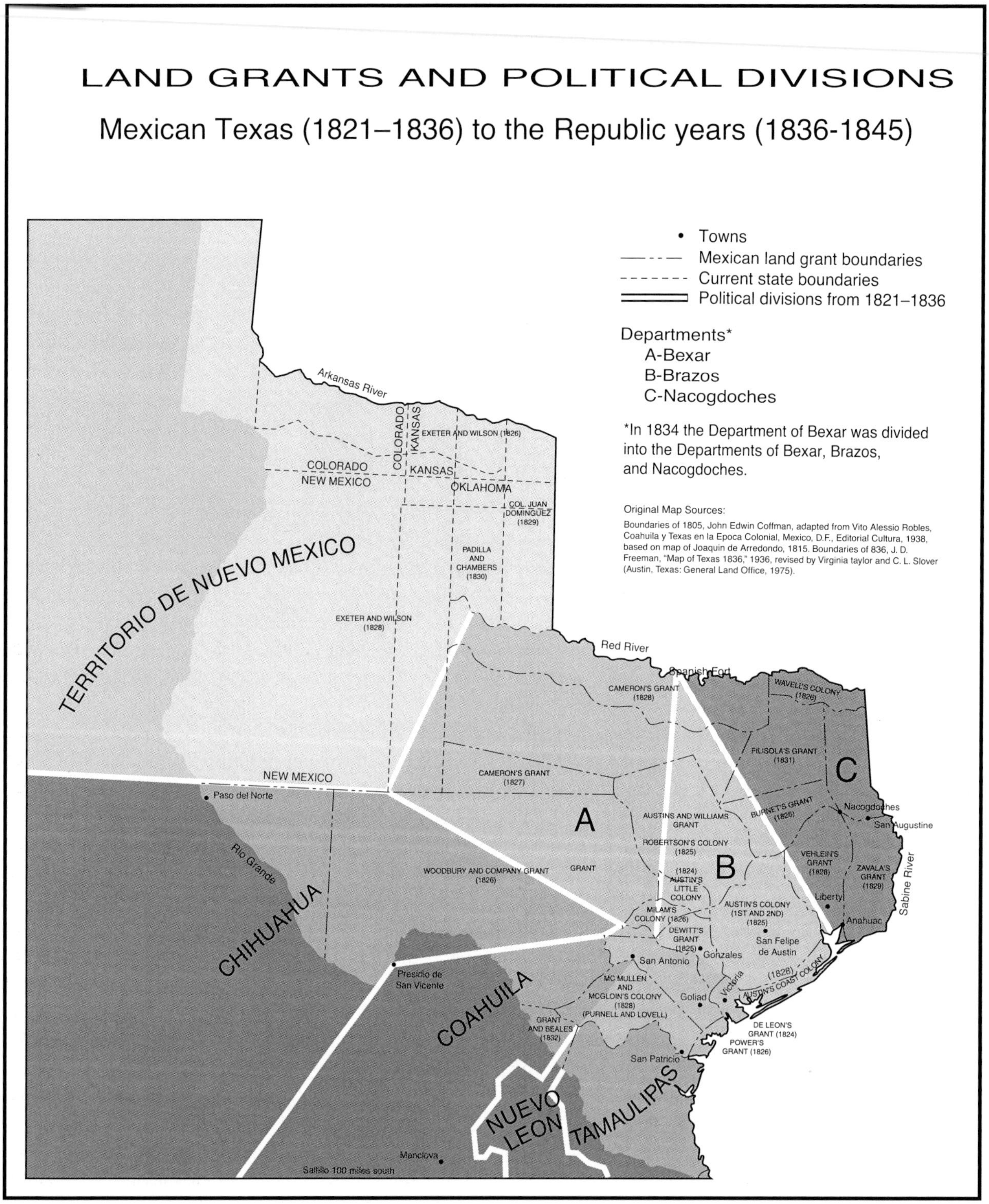

FIGURE 3-9 Republic of Texas
Standard depiction of the Republic boundaries with Empresario divisions. Previously published in the *Atlas of Texas*.

FIGURE 3-10 San Jacinto Monument
San Jacinto Battleground State Historic Site, Harris County.
Photo courtesy of Texas Department of Transportation.

statehood. The Mexican government hoped to focus development on the department's respective capitals as well as any new ports, such as Anahuac, and they intended to garrison troops lead by experienced Mex-Indian leaders.

Immediately, the three departments had very different characteristics. Bexar was the most Hispanic with longest tenured places of San Antonio and Goliad. San Antonio was the official capital for the Bexar department as well as the de facto main place for all of Texas. Spanish ranching complex was successfully introduced and Spanish-Mexican cultural landscapes were common. Brazoria was centered on the Austin Colony with San Felipe as its capital. This department had an Anglo majority, but it was very stable because of the screening of migrants by Austin and others. Eventually, Southern planters attracted to the bottomlands with river transportation and later distinct European groups would migrate into this part of the state. Finally, Nacogdoches was the capital of the third department of the same name. This department was the most Anglo and it was the initial part of the state that most migrants from the United States were settling in. With proximity, it had social connections and economic activities directed to the American South including New Orleans. In addition, it absorbed Native American groups and eventually African Americans when slave-owning planters arrived. Technically, Nacogdoches administered the Red River area, but actual Mexican presence and control were very minimal.

Mexican Texas was short lived (1821–1836) but numerous developments occur in this time span. High population growth rates begin. The absolute number of residents in 1836 is estimated around 30,000, which is a large percentage change from the Spanish period. However, the composition of settlers radically changes toward an Anglo-American bent with most grants arriving from the U.S. South. Likewise local government and property distribution begins in earnest. Most significantly, 3500 land titles are granted during Mexican Texas. This means there are numerous landholders (primarily families) who have a vested interest in the future. If the Spanish laid out the initial framework for Texas, the Mexicans started to fill it up with inhabitants. Furthermore, they had to make adjustments to governing that most new arrivals were relatively accommodating with and kept afterward.

1836

1836 becomes the transformative year of the "Texas Revolution" with the recognized transfer of Mexican sovereignty to the Republic of Texas. The events of 1836 are well-known but worth further mention. These events were preceded by numerous others such as the Consultations (political developments) and the Gonzales incident (other potential/actual acts of violence). There were three quick battles in the spring of 1836 that obviously dictated the geopolitical outcome. The conflict also propelled a major disruption of people (Scrap) and patterns as some places like San Felipe never really regained their significance.

Declaration of Independence in Washington on the Brazos, March 2nd. The political evolution of Texas is made clear with the formal break from Mexican sovereignty.

Alamo: Mexican army's siege of the Valero Mission being used as a fortified garrison by Texans and American volunteers: Travis, Crockett, Bowie, etc.

Goliad: The surrender and execution of Fannin's troops by the Mexican army.

San Jacinto, April 21st. Texas troops led by Houston defeat the Mexican army and subsequently captured the Mexican President and Commander-in-Chief Santa Anna.

REPUBLICAN TEXAS

The Republic of Texas existed for only a decade, but it had a tremendous impact on the eventual geographies of Texas. Perhaps the crucial dimension was the idea of empire because it shattered the idea of a contained Texas. Before the Republic, Texas was an ambiguous frontier region that was peripheral in every sense of the word. Now as an independent entity, Texas could imagine itself and project its influence beyond the coastal plain. The previous U.S.–Mexican border was already defined and the Rio Grande was claimed as the remaining southern and western boundaries. This "shape" of Texas was officially claimed until 1850 and it is cartographically ingrained into our collective memory.

During the Republic, two images of Texas emerge. One is a pragmatic image of Texas that eventually joins the United States. Considering the vast majority of inhabitants and new migrants were coming from the southern United States, it is no surprise that this was the end result. However, the second image of Texas was of an empire: a third independent entity between the United States and Mexico that would eventually extend to the Pacific. Significant number of early leaders thought in such grand terms that can only be compared to imperial aspirations and actions. The most significant statement of empire was the selection of Austin as capital. Austin exemplifies the concept of a forward capital because it was on the edge of settlement in 1839, but it symbolized the future. For the Texas as empire intellectuals, Austin was imagined as the future crossroads: the intersection of North–South (United States and Mexico) and of Coastal–Inland (Gulf of Mexico and Santa Fe) interactions. Although Texas did become a state, this imperial impulse contributed to the cultural dimension, and perhaps this is where boldness became synonymous with Texas.

Having ownership of all the public lands, the Republic could impact settlement in a dramatic way. They gave away land to veterans and survivors, and they set out to sell land to new migrants, which they sold at slightly lower prices and larger plots than the U.S. government. This greatly accelerated the population of Texas, which reached 150,000 by 1845. Much of this settlement was in-fill of previously settled areas, but overall, the settlement frontier was reaching the Balcones Escarpment (Castroville, New Braunfels, San Marcos, Austin, and Waco). Soon thereafter contact with the Plains Indians would inhibit further expansion except in isolated places where locals negotiated deals with indigenous groups such as Fredericksburg or as gateway towns linked to military forts. Although an expansive Texas is/was imaginable, most of the state becomes effectively settled only after 1845.

STATEHOOD

Even if Texas' admission to the United States was inevitable, domestic and even international concerns altered the exact timing. An annexation agreement was negotiated in 1844, but it wasn't able to get through Congress. That summer, a presidential candidate named James Polk ran for office with the slogan "Oregon and Texas," meaning that if elected he would immediately do what was necessary for these two areas to formally become American. Polk went on to win his party's nomination and the general election. Even before he was sworn in as president, Congress moved to annex Texas. Texas was admitted as the 28th state on December 29, 1845.

Mexico's displeasure with Texas annexation was well-known, and British mediation was not working. When U.S. troops appeared south of the Nueces River, Mexico declared a "defensive" war. The first skirmishes occurred along the Rio Grande, but much of the war was fought in Mexico proper with the final battle being in Mexico City itself. The Treaty of Guadalupe-Hildago set the U.S.–Mexico boundary with Mexico ceding their entire northern frontier and recognition of the Rio Grande as the border with Texas. For the United States, the war was Manifest Destiny in action with a military conquest that expanded America to the Pacific. The treaty defined the southwestern quadrant of the United States.

Now that the U.S.–Mexico border was defined, the U.S. and Texas governments had to negotiate the exact boundaries of the state. With U.S. troops throughout the region—occupying Santa Fe—and along the border, Texas' claim to New Mexico was appearing tenuous at best. The agreement known as the Compromise of 1850 defined the final three segments of the border: Northern Panhandle, Western Panhandle, and 32° N latitude lines, which produced the current, recognizable shape of Texas. Analysts often describe the compromise as the feds got their map and the Texans got their money. Texas received a cash payment that allowed the state to finally pay off their Republic-era debts.

Statehood provided Texas with exact boundaries, but it did not settle those unsettled areas. It took another half century or so for enough people to organize the whole state into the 254 counties we know today. The trends were already in place: increasing numbers of people and a westward moving settlement frontier. Although large scale settlement of West Texas and the Panhandle had to wait until after the Civil War, it happens quickly when a concerted effort to subdue the Comanche began. Coinciding with settlement are technological changes that impact transportation patterns and economic activities. It is safe to say, at this time Texas is securely part of America.

"Music as Cultural Voice"

FIGURE 4-1 Austin and Live Music
Sixth Street in Downtown Austin, Travis County. Photo by Author.

4
Cultural Diversity

The cultural geography of Texas is interesting because under the ten-gallon guise of similarity there is much diversity. Even the stereotype of Texans is interesting because it tells much about who we are at various geographical scales. Why has one particular image, namely the cowboy, become the dominant stereotype? At a national scale, many Americans associate Texas with ranching and cowboys, which is a relatively accurate stereotype compared to themselves. It is this particular image of Texas that diffuses around the world. However, at a local scale, there are numerous alternatives such as farmers, lumberjacks, and fishermen not to mention the array of modern professions. While most people would tend to associate cowboys with the rural white population, the past and present exemplify a multicultural dimension.

Literally anyone can wear a cowboy hat, and everyone can choose their actual relationship to the stereotype. Chapter Four explores the cultural diversity beneath the cowboy hat. As it turns out, there are bonnets, berets, and bandanas as well as a few fedoras and macaronis. After elaborating on the concept of culture, we detail the large migration movements to Texas. Then we turn our attention to two special elements of culture: language and religion.

DEFINING CULTURE

Culture is both a simple notion and a complicated idea. One often describes self in cultural terms, and many people see place through a cultural lens. In effect, culture is who we are as well as where we are. Yet, in the big realm of ideas (social theory), there's much debate about the existence let alone the definition of culture. One contemporary definition of culture is the **"shared system of meanings"** that relate to the beliefs, behaviors, and material belongings. This definition allows for different-sized groups (two to infinity) and overlapping membership (one can be both Texan and American) while only alluding to its social construction.

CULTURE

Culture is a debated topic in academia because of its many nuances of meaning combined with the self-evident significance that people give it. On one hand, the all encompassing nature of culture defies a simple definition. Culture feels like an omnipresent phenomenon that dictates our speech, actions, and morals. While looking at other people, we assume those differences are "cultural"; for example, French drink quality wines, Polynesians are welcoming of strangers, and Aborigines are in tune with their environment. At some level, we could imagine someone saying it's their culture as they try to explain these qualities. On the other hand, most people accept culture and strongly identify with these labels. We can imagine someone standing tall, boastfully saying I am Scottish, Quebecois, or Texan. But what does that actually mean? Cultural determinism is when people justify their actions by declaring their identity, or in effect, my culture made me do it.

Culture is better conceptualized as a process than an absolute judgment. When archaeologists interpret newly discovered artifacts, they try their best to differentiate the human past into spatial areas and temporal periods of culture. When high school teenagers segregate into different social groups, they coalesce around similar slang, dress, and labels; similar sub-cultural groups are common with adults in America based on what one does for work or entertainment. Both of these examples show that culture is about differentiating one group from another. Differentiation is at the center of culture. Furthermore, much of the social construction of difference is oppositional. Like the individual's development of self from other, culture is a process of creating "us" with the help of defining "them." It is argued that this "othering" has a geographical aspect which manifests itself as regions. The most discussed example comes from Edward Said's *Orientalism,* whereas European identity was constructed in oppositional terms with the Middle East/Orient.

As a geographer, I would contend that the construction of culture is grounded in places and regions. Moreover, the differentiation is experienced in lived in places where ordinary people negotiate who they are and what this place means with their fellow citizens. My definition of culture is the collective negotiation of place that differentiates it from the rest of the world. This negotiation of place process defines both identity (who we are) and territory (where we are). Miles Richardson would say "culture in its place"; not only does every place have its own culture, every culture is grounded in the reality of some place.

Once we accept the general idea of culture, then we can also start to think of groups and regions associated with "culture." A culture group is a collection of people with shared systems of meanings such as speaking a unique dialect or worshiping together. In the context of Texas, the culture groups are consciously aware of who they are and where they are arriving from. Moreover, they know how they differ from others already in Texas. As these different groups settled in Texas, cultural patterns developed that have some clear regional dimensions.

MIGRANT ORIGINS

Logically, the original migrants to Texas were Native Americans. We do know about the more recent movements such as the Comanche and Cherokee that influenced Texas' historical geography. Additionally, there is a modern dimension to Indian populations in Texas, which will be covered in Part Four. In this section, the culture groups are discussed with an eye toward the recent past and their areas of emigration: Anglo-Americans, direct Europeans, Mexican-Americans, and African-Americans.

Anglo-American (White)

Anglo-Americans are the largest group in aggregate, but there is quite a lot of internal variety. The key point is that this group of migrants emigrated from the pre-existing United States. Therefore as a group they had adapted many cultural traits from living in the United States, and then they diffused those ideas and attitudes to Texas. Fundamentally, they were Americans. Those that arrived before 1836 could be described as something more than just "frontier" because they were in the forefront of American expansion to the point that they actively entered Spanish and Mexican controlled areas. It is one off those rare circumstances when the ideology of land ownership, individual renewalism, and personal opportunity were greater than actual citizenship or reliance on government. Those that arrived after the Republic years were clearly part of the wider, westward advancing American frontier, yet they also expanded the sectionalism of southern slave states.

When cultural geographers and historians look at the United States, they generally see an East to West pattern with the original 13 English colonies on the Atlantic and the eventual expansion to the Pacific. Manifest Destiny and the "Frontier" are two of the big ideas that still resonate because of the legacy in the way we think about resources and foreign policy for example. Of course, this East–West story doesn't adequately explain the American Southwest (Texas to California). Yet, it helps us understand the variations of "American" culture that diffuse to Texas. Those 13 colonies were different from one another based on their environments, the types of emigrants they attracted, and the nature of royal charters which spelt out the economic and political purpose. Typically, the colonies are generalized into New England, Mid-Atlantic, and Southern. Furthermore, some scholars think of these three colonial groupings as cultural hearths for American sub-cultures. With enough time in relative isolation from one another, cultural features such as dialect, (vernacular) construction, and acceptance of slavery develop.

Out of these hearths, those cultural variations diffuse westward along specific pathways with American expansion. The New England subculture diffuses to the Upper Midwest/Great Lakes and beyond in an insular fashion (cities) to the west coast. Boston is the archetypical city for New England and with Philadelphia one of the models for the northern half of the United States. The Southern subculture originates in the Virginia Tidewaters, and it expands into Carolinas with the success of plantations along the coastal plain. Eventually, this subculture diffuses westward with migration along the Gulf of Mexico to eastern and coastal Texas, yet it must interact with French and Spanish cultural elements to some degree. Scholars often refer to this subculture as Lower South because it is a lowlands environment that planters utilized for slave-based, intensive agriculture. The Lower South develops a unique urban environment and model for southern cities based in part on Charleston and New Orleans.

In between the New England and Southern subcultures was the Mid-Atlantic subculture, which originated in southeastern Pennsylvania along Delaware Bay. Scholars associate this subculture with broad themes of America such as ethnic mixing and an amalgamation of ideas and practices, which becomes the basis for all-American and generic Middle America. Not only can one list the groups, Germans, Swedes,

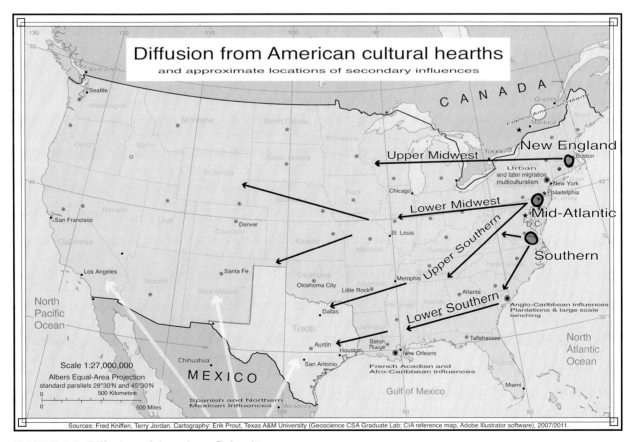

FIGURE 4-2 Diffusion of American Subcultures

Scots-Irish, English, etc. but the different knowledges such as farming, construction, and governance that each group diffused with to America can be researched. The Mid-Atlantic subculture diffuses wide and far in its movement west and clearly becomes the largest region in terms of area. There were different pathways that are often significant enough to articulate: Lower Midwest and Upper South. Lower Midwesterners settled those areas North of the Ohio River and eventually beyond the Mississippi River and into the Missouri River valley. Upper Southerners initially migrated in a southwesterly direction along the Shenandoah Valley where yeomen from the Virginian and Carolina lowlands could move inland and theoretically reinforce Southern identity. Then they crossed into Kentucky and Tennessee through gaps in the Appalachian Mountains. In the lower Mississippi River basin, Upper and Lower Southerners crossover as planters expand upstream along the fertile river basin and back-country highlanders occupy the Ozarks/Oauchita of Arkansas and eventually North and Central Texas.

From a Texas perspective, the two dominant flows of Anglo-Americans were from the Mid-Atlantic and Southern cultural hearths. More specifically, the earliest Americans were from the Upper South. Historical figures such as Austin, Houston, and Crockett all hailed from the upland South with Tennessee as the primary point of previous residency. These upper Southerners traveled overland exploring and appropriating good farm land on the coastal plain; early explorers gave way to more organized townships and communities. Many of these upper Southerners crossed into Texas via Arkansas, and sequentially they settled Red River Valley, North Texas (Dallas), Cross Timbers, and much of West Texas.

From the Lower South a movement of planters settled along the bottomlands of river valleys and in some cases overwhelming the earlier upland settlers. The lower Southerners arrived by boat or by wagon often in very coordinated migrations with possessions that included their most important asset, slaves. The origins for lower Southerners vary across the slave-owning South especially the Gulf Coast states. The potential advantage of proximity for Louisiana planters is noticeable along the Gulf Coast, but overall the largest number was from Alabama. Lower Southerners eventually settled much of East Texas and the lower river valleys along the coast, and they initiated the need for gateway cities such as Galveston for transportation of goods and services.

In addition to upper and lower Southerners, lower Midwesterners settled a distinct part of Texas: the northern Panhandle. As part of a broader migration from the lower Midwest, Anglo-Americans settle on the northern and central sections of the Great Plains after the threat of Indians diminish in the later decades of the 19th century. Further south, it was second- and third-generation Texans (mostly upper Southerners) that occupy the southern Great Plains. After initial settlement, there was reinforcement of Mid-Atlantic culture as strong trade and transportation links develop between Texas cities (especially Dallas) and Midwest cities such as St. Louis, Kansas City, etc. While not occupying a distinct region, individuals from the New England subculture also migrated to Texas. These migrants primarily resided in the fast growing cities and gateway locations such as Houston, Dallas, and Galveston.

Direct European (White)

While almost all Anglo-Americans that arrived in Texas ultimately emigrated from Europe, Texas was a direct destination for European migrants. Time has a way of blurring the distinctions between Anglo and other Europeans; however, the initial cultural differences of settlements in Texas were tremendous. Firstly, European immigration to Texas occurs in a relatively organized fashion with reputable land agents and shipping companies. The roots of this movement date back to the Empresarios and Republic land sales where Europe is seen as a potential source of settlers. Some of the first migrants become intermediaries for future migrants, and support groups are created to assist and facilitate migrants. Secondly, Europeans had the ability to migrate as groups. For example, a whole boat load of migrants may come from the same region of Europe and sometimes even the same town. Then, these groups could settle together when they arrived in Texas, and they often tried to replicate their European cultural landscapes. Because they didn't have to integrate with Anglo or Mexican communities, they could retain their language and religious beliefs for many generations. Thirdly, there was a geographical dimension to this

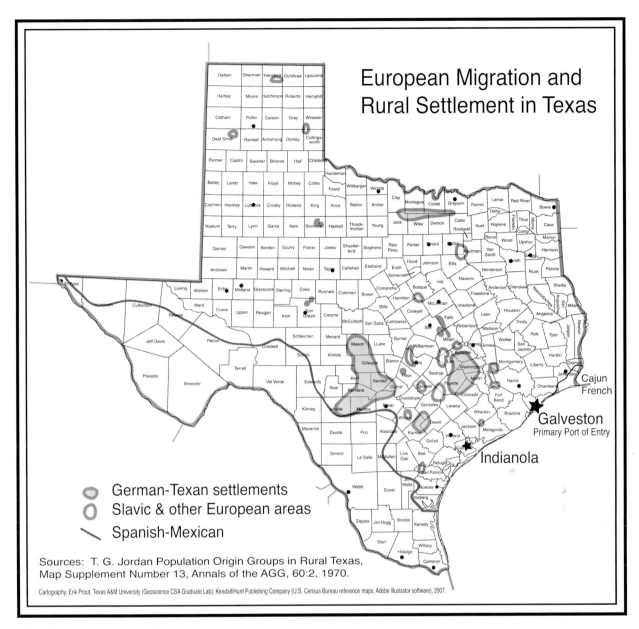

FIGURE 4-3 European Texas

direct European migration. At the local scale, communities of a single place or closely spaced multiple towns had a distinct Euro-Texan identity. At a regional scale, German and Czech migrations were large enough to perceive as coherent regions consisting of multiple counties. Parts of Texas such as Hill Country have a strong European association. At a state scale, European migration was focused on the middle coastal plain between the ports of entry at Galveston and Indianola. Yet individual European settlements are all over the state and often intermixed instead of a discernable segregated pattern.

The largest group in terms of population was the German-Americans. The push factors for German migration include numerous events in Central Europe such as the 1848 Revolution, rapid industrialization/population growth, and Bismarck's unification of Germany. The allures of inexpensive, productive land together with ethnic connections were strong pull factors. As early as the 1850 census, the number of German-Texans exceeded the number of Mexican-Texans. Most German immigrants settled along a beltway of counties west of Houston, and it is here that the first German community of Industry was founded.

FIGURE 4-4 Wendish Community, Serbin
Texas Wendish Cultural Museum, Lee County. Photo by Author

Because this area was also settled by Anglos who had adapted to the South, they adapted similar landscapes that were well-suited to the environment. However, those areas where Germans migrated to in Hill Country became majority German, and they transformed the landscape in ways that still appear distinct today. San Antonio and New Braunfels became important central places and launching points for Hill Country settlement, and places like Fredericksburg and Kerrville become central locations for German communities farther inland.

The second largest direct European group in Texas is the Czechs. Czech communities can be found in a similar pattern to the Germans on the coastal plain and like other Europeans diffused throughout Texas except perhaps the German-dominated Hill Country. Most Czech-Texans emigrated from Moravia (East), whereas the others came from Bohemia (Prague/West). Other Slavic speaking groups immigrated to Texas besides the Czechs; they include the Slovaks, Poles, and Sorbs. The Slovaks and Poles like the Czechs emigrated out of the old Austro-Hungarian Empire, and most of them migrated through Germany using much of the same infrastructure the Germans were using to emigrate. The Sorbs are a very small minority group in eastern Germany, and they also are known by the ethnic label Wends.

Scattered around Texas, there are numerous individual communities associated with European migration. As the absolute number of settlers is low, it is hard to group them together into meaningful categories. We can classify the northern European migration with the term Scandinavian: Swedes, Danes, and Norwegians migrated in small numbers and created separate settlements like other European groups. Mediterranean Europe also emigrated to the United States and in Texas we find the occasional Italian migration in rural areas. Irish migration to Texas has the longest historical dimension when we remember that San Patricio and Refugio predate 1836. Later migration of Irish and Italians impacted select urban areas throughout the United States. Alsace, the contested border province between Germany and France, provided migrants for the Castroville settlement. Another category of migration is ideological in that the immigrants were "escaping" and starting anew. Religious groups like the Mennonites, Socialist/utopian groups, and secular German "Freethinkers" are all examples found in Texas. We also have a spill-over settlement pattern along the lower Sabine/Louisiana border that can be seen as having a European dimension. French-speaking Cajuns, who themselves were diasporic migrants from Canada, expanded westward from their core areas into the Upper Coast/Southeast Texas. In summary, European migration in a direct way (Germans, Czechs, etc.) and in indirect ways (Americans, Cajuns, and Mexicans) had an enormous influence on the demographic structure of Texas. During the census counts of the later half of the 19th century, the official counts of the foreign born population were under 10 percent. The cumulative impact of second and third generations that retained their parent's languages is what implanted European culture and landscape in Texas.

Mexican-American (Hispanic)

The first European settlers and settlements in Texas were of course Spanish. In many ways this migration should be conceptualized as a secondary movement and not a direct migration. The Spanish expanded into Texas through Northern Mexico, and in Mexico the demographic dynamic was very different from the English based processes in North America. The Spanish intermingled with Native Americans which included marriage and children as part of a larger effort to integrate the indigenous population as Spanish subjects. After a few generations, the variations of mixed were numerous and eventually this category, Mestizo, becomes the statistical majority. Modern Mexico internalizes and celebrates this hybridity between Euro-Spanish and indigenous Aztec/Mayan as a positive thing. However, there remain extreme socioeconomic consequences with being Spanish (good) and Indigo (bad). Therefore, one should see the first post-Columbian settlements as Mexican. The complex human variation that is Mexico arrives with the Spanish settlement. While there were a few direct and indirect Spanish migrants to Texas, the majority were Mestizo. In addition, the diffusion of Spanish culture is also mediated by the pathways, adaptations, and modifications that occur in Mexico. Landscape preferences and modifications that diffuse to Texas are very similar to those in northern Mexico.

TABLE 4-1 Foreign-born population

Year	Foreign born	Percent Foreign
1850	17,681	8.3
1860	43,422	7.2
1870	62,411	7.6
1880	114,616	7.2
1890	152,956	6.8
1900	179,357	5.9
1910	241,938	6.2
1920	363,832	7.8
1930	362,287	6.2
1940	235,528	3.7
1950	277,515	3.6
1960	298,791	3.1
1970	309,772	2.8
1980	856,213	6.0
1990	1,524,436	9.0
2000	2,889,642	13.9

Statewide average (1850–2000) is 6.6% foreign born.
Source: U.S. Census Bureau, 2000

Those areas that were successfully settled by Mexican migrants in 1836 are primarily in South Texas. The San Antonio River valley was the most densely settled area. San Antonio was the largest place with numerous missions, a presidio, and a budding commercial town. The Canary Islanders that received the 1731 City Charter are also one of the few examples of direct Spanish migration, and even then the Canary Islands are not adjacent to the Iberian Peninsula. Downstream from San Antonio the La Bahia missions were another nucleus of Mexican settlement in the river valley. North of the San Antonio River, Mexican settlement was sparse and intermixed with Anglos at the main places for Empresarios and government operations. Mexican officials and entrepreneurs were present around the state, but it did not necessarily constitute effective settlement. The remaining area of Mexican settlement was along the Rio Grande, which had three concentrations of missionary activity: El Paso, Presidio, and the lower Valley. None of this settlement was conceived as being part of Texas to Mexican officials and settlers until after the Mexican-American War (Treaty of Hidalgo-Guadalupe borders). The (lower Rio Grande) Valley was being settled on both sides of the river as the northern part of Tamaulipas. In addition, various river crossing points were developing small settlements such as Laredo and Eagle Pass. Further upstream, El Paso was still seen as the northern sister city of Chihuahua.

After 1836, Mexican presence in Texas diminished considerably. In some instances, there was harsh treatment after the wars which facilitated movement southward, and then later the Republic and occasional U.S. resistance to new Mexican migration reduced the numbers. South Texas was the exception as Mexican culture endured until further immigration from Mexico resumed. Immigration to Texas becomes very cyclical with ups and downs of both push and pull factors. Push factors out of Mexico are very strong when political turmoil is high or economic conditions are low. Those periods when quality of life is stable or improving translate into minimal push factors. Pull factors to Texas (and America) are very strong with economic growth and prosperity in Texas. When quality of life conditions go down in Texas, tolerance of migrants does as well, hence lower pull factors because of the social and legal restrictions. One should remember that there are times when Texas and the United States are not in economic sync; for example, growth in Texas generally outpaces the rest of America and Texas had boom/bust cycles with energy. Those unique times when there was both strong push (from Mexico) and pull (to Texas/America) factors resulted in very high migration rates.

The first big wave of Mexican migration coincides with an agro-economic innovation in the early 1900s. Market gardening or truck farming develops with the advent of automobiles; the idea is to invest heavily in intensive production of valuable crops and drive them quickly to populated markets. The Valley was one such location in the United States where the soil, water (irrigation), and annual temperature were ideal for year around intensive agriculture. Once transportation and labor were in place, the boom began. The labor force came from adjacent Mexico, not only in Texas but also California. In South Texas, the Mexican culture was still present and ready to absorb this influx of new people. Some of new immigrants stayed in South Texas, while others moved onto other employment opportunities. A pattern develops that one observes today with South Texas serving as a launching point for continued migration to the central and eastern United States.

As a broad category, the Census does not use Mexican. Instead they use the term Hispanic, which properly includes the other origins for migration in Latin America. In the United States as a whole the Hispanic category is still dominated by historical and contemporary migration from Mexico. Nevertheless, large populations of Puerto Ricans, Cubans, and Dominicans—generally along the East Coast—are noteworthy. Increasingly migrants from Central America are becoming a significant part of the Hispanic category, which is geographically interesting because they are using some of the same pathways as traditional Mexican migration to America. As most of us are aware, migration from Mexico to Texas has both legal and illegal dimensions. Almost all of the Latin American migration through Mexico as well as a significant percentage of Mexican migration is undocumented, which is by definition illegal. Therefore, we can only estimate the true movement across the border and the dimension of Hispanic-America. Logically, this migration even if partially undocumented and undercounted is reinforcing the previous Spanish/Mexican cultural diffusion.

African-American (Black)

The last category is a unique group to discuss because African-Americans are the only culture group in Texas that we associate with involuntary migration. Involuntary migration is the movement of people without voluntary consent. Of course there are instances when people are coerced and pressured to migrate especially when the push factors are extreme such as war and famine. The forced movement of African-Americans as slaves to Texas constitutes a classic example of involuntary migration. In the American South, slavery was associated primarily with the individual agricultural plantations rather than societal works projects or elitist creature-comforts. Plantations were more prevalent in the tropics where high value crops such as sugar could be grown, and the long-term profitability of these agricultural endeavors rested on an inexpensive labor force. Plantations in the South were household economic actors that could and did fail because the plantation agricultural model was tenuous at that latitude. American plantations originated on the East Coast and they began with tobacco, rice, and indigo crops. Eventually, the crop of choice became cotton, and cotton plantations expanded along the Gulf Coast to Texas. Most slaves from Africa arrived in the Caribbean, and then the involuntary migration to the United States was a secondary movement that funneled through a few seaports.

African-American migration to Texas coincides with the Lower Southerners. As these planters moved their plantations to Texas, they brought their slaves to the same destinations. Therefore, the initial geographical pattern of African-Americans relates directly to the distribution of plantations, which were primarily in East Texas and the Gulf Coast along navigable rivers. Ironically, the birth locations of slaves differ slightly from planters. By the time plantations are being created in Texas, the trans-Atlantic slave trade was in decline and the importation of new slaves had been banned. Slaves had to be purchased from previously existing plantations or slave markets. As a percentage, more African-Americans in Texas were born in Virginia, Georgia, and the Carolinas than Lower Southerners. One could speculate that the slaves had more New World agricultural experiences than their owners including knowledge of cotton production and ranching tasks for example.

Illegal diffusion of slavery began during the Mexican Texas era when slave-owning Southerners migrated to Texas. One of the consequences of the Texas Revolution was that slavery became legal, and the pace of migration for Lower Southerners and their slaves increases. African-Americans became a larger and larger percentage of the population between 1845 and 1865, which means their growth rate that includes both natural increase and migration exceeded all other groups. During the 1860 Census, slaves made up nearly 30 percent of the population. In some counties, slaves were statistically the majority. After the Civil War, African-American migration to Texas becomes zero. Without immigration, the percentage of the population declines despite the fact that the absolute population is increasing. That absolute number is 2.4 million, which means Texas has the second largest African-American population of the 50 states.

Initially, the geographic pattern to African-American populations remained very stable as many stayed on as sharecroppers. Over time the African-American population becomes more segregated as they are relegated to their own rural settlements and to specific parts of larger towns. In the 20th century, a large-scale rural to urban migration occurs in America, and for rural African-Americans in the South the movement has a regional quality toward the North. In Texas, we see some of the same types of migrations, but the fast growing cities attract rural African-Americans as well as others. Currently, the geographical pattern for African-Americans includes both urban places and those traditional rural areas. However, no county is more than 25 percent African-American and only a few counties are less Hispanic than African-American.

LANGUAGES

Currently, English and Spanish are the two most spoken languages in Texas; however, there are numerous other languages being spoken. With Texas cities attracting people from all over the world, it is unsurprisingly that some of the world's linguistic diversity is found there. In the past, linguistic plurality could be experienced in rural Texas because ethno-cultural communities kept to themselves and preserved their language as best they could. One could say that Texas has always been linguistically diverse, has had those special people who could communicate and translate between groups, and it still has a dynamic linguistic situation.

Language is an integral part of human culture. Language is the means by which we not only negotiate our daily needs but construct our identities at various geographical scales. Using the previous definition of culture, "language" is the shared system of meanings associated with sounds and symbols. For most of human history, language was front and center for oral culture. Over the last 6,000 to 8,000 years, writing diffuses to the majority of humanity but not necessarily all languages. With the rise of written culture, modern humans conceive of language as both the oral sounds of speech and the visual symbols of writing. Most people closely identify with the labels for their language such as English, Spanish, and German, and it is common to link languages to ethnic groups. For the most part, linking Americans to English and Mexicans to Spanish is correct, but it is not correct to assume that all immigrants from Germany spoke German and all Hispanics speak Spanish. In addition, languages have dialects or small variations that reveal differences between the speakers. Geography is integral in linguistic variation because distance and barriers are often the principal factors, which coincides with place and region. The term "vernacular" conveys both regional dialects and those variations based on social differentiation.

Chronologically, the first languages spoken in Texas do not exist. The over 10,000 years of human interaction had evolved by the time of Columbus. In general, very little of the indigenous linguistic past survives, and what does comes from the Native Americans that actually interacted with the earliest Europeans. One of the legacies is place-names: Waco, Nacogdoches, and Neches River exemplify this. Of course the Hasinai-Caddo word that evolves into "Texas" is significant. As European colonial power asserts itself in the New World, their languages do so as well. Spanish, English, Portuguese, and French (in numerical order) diffuse and expand over large areas of the Americas. For Texas, the migrants speaking two colonial languages, English and Spanish, persevered the best.

English

The English language arrived in Texas with Anglo-American and African-American immigrants. Literally starting with Austin, English speakers settle the middle and eastern parts of the coastal plain. By the 1830s, English was already the most spoken language in Texas despite Spanish being the official language. After the Revolution, English becomes the de facto official language, and Anglos expand into the southern and central parts of the state. Finally, English speakers settle the western side of the state in the later half of the 19th century. Being part of the United States, English in Texas is reinforced by those national interactions.

Interestingly to cultural geographers, the English that diffuses to Texas has strong variations. Those three American subcultures discussed earlier have definite linguistic variations. Therefore, we can trace their pathways to Texas, track the modifications, and map the patterns of language. The four most common historical dialects are Northern, Midwestern, Hill Southern, and Plantation Southern. The Midwestern and Hill Southern originate together with the Mid-Atlantic subculture with the other two respectively to the New England and Southern hearths. In addition, African-Americans provide an interesting dimension to the relocation diffusion and evolution of English in Texas, which will spatially overlap with Lower Southerners.

The first vernacular dialect of English arrives with Upper Southerners and is labeled Hill Southern. Explorers and settlers speaking Hill Southern got first opportunity to name places and environmental features. For example, terms such as hollow, gap, and fork in Texas derive from Hill Southern, and they replicated many place names from their experiences in the Upland South (Tennessee and Arkansas). Hill Southern speakers also had the earlier chances to interact with Mexican settlers, so using their translations or pronunciations of Spanish were more likely to become the standardized English. Furthermore, as these Upper Southerners expanded into Central and West Texas, they seemed accepting of Spanish loan words. Spanish words like mesa and rodeo are used seamlessly in Hill Southern. Another English dialect closely related to Hill Southern, Midwestern, diffuses to Texas. With the Lower Midwesterners, this relatively small linguistic region establishes itself in the northern Panhandle.

TABLE 4-2 Languages in Texas, 2005

Language	Percentage of Total
English only	66.4
Other languages	33.6
	Percentage of Other
Spanish	86.5
Vietnamese	2.0
Chinese	1.6
French	0.9
German	0.9
Tagalog	0.8
Urdu	0.7
Korean	0.7
Hindi	0.5
Arabic	0.5
Others under 0.5%	5.0

Source: U.S. Census Bureau, Languages Spoken at Home, 2005 (based on population 5 years and over of 20,403,745).

As Lower Southerners and African-Americans migrate together to Texas, Plantation Southern diffuses to the eastern and coastal parts of the state. Plantation Southern uses the terms gully and bayou for environmental features. In those areas with large movements, Plantation Southern becomes the dominant English dialect. On linguistic maps, there is a broad zone of transition between Southern dialects. Furthermore, it appears as if Plantation does not necessarily overwhelm Hill dialect as much as migration alone would suggest. Plantation Southern is often seen as the language of the stereotyped southern planters, but it was also spoken (albeit differently) and influenced by the slaves. In reality, it is a product of black/white coexistence that continues into the present. The colloquial term "Black English" is heard as part of the discussion, which has been used pejoratively; contemporary linguists call it African-American Vernacular English (AAVE). Linguists have documented the uniqueness of this vernacular and discovered that some aspects are actually common to West African languages. Despite involuntary migration and force language conversion, linguistic as well as other cultural elements survived.

Spanish

Spanish diffused to Texas earlier than English, so it was the first colonial language to implant itself. Yet, English speakers outgrew Spanish speakers when Anglo-Americans became the dominant source of settlers. Spanish survived in Mexican-Texas families and communities, which had a strong regional pattern in South Texas. The earlier arrival of Spanish speakers means that they had the first chance to interact with indigenous peoples and create place-names. Spaniards named all the major rivers and topographic features. For example, Sabine is the Spanish word for cypress, and the Neches were a Caddo tribe that the Spanish adopted as a place-name. Americans often shortened or translated Spanish names; Brazos derived from the "Rio de los Brazos de Dio," and Trinidad became Trinity.

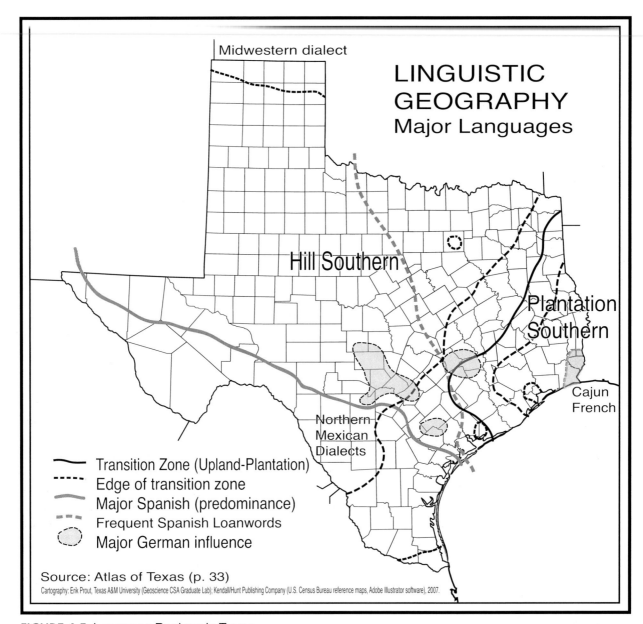

FIGURE 4-5 Language Regions in Texas

Like English, it would be foolish to think everyone spoke an identical Spanish dialect. Besides the social distinction of Spanish/Mestizo/Indian, there are Spanish dialects in Spain that diffuse to the New World. In Mexico, there are significant regional variations that include not only language but other cultural elements such as ranching, cuisine, and patron saints. Therefore, the historical and contemporary migrations of Mexicans to the United States reflect these geographies. Proximity is important in this instance as migration to Texas has been partially from and primarily through northeastern Mexico. More recent migration is solidifying South Texas as a Spanish speaking/Mexican cultural region, which many elude to the label Tejano. Tejano is generally understood to be Mexican-Texas. There are some interesting linguistic situations exemplified very well in South Texas: "Border English" and "Spanglish." When two language communities coexist together, it is common for simplified pidgins to develop and equally likely that speakers of both languages will mix words/ideas of one into the other.

Other Languages

English and Spanish are the two most spoken languages, but a quite a few others had and are having an impact on Texas. For ease of order, there are those languages associated with Direct European migration and those that are of a contemporary migration. Of the historical ones, German stands out for its number of speakers and the associated landscape impact. During the middle decades of the 19th century, German was the second most spoken language in Texas. There were German language schools, newspapers, church sermons, and everything else one would consider normal. Unfortunately, the German-American experience was made more difficult because of the World Wars. German-speaking communities actively gave up their public use and displays of German as an act of patriotism and/or fear of the consequences of not doing such. For the other European languages, Slavic Czech, Slovak, Polish, and Wendish as well as Irish Gaelic and Cajun French, there wasn't such a dramatic event. They mostly experienced the slow transition of assimilation into the dominant culture that was the white English-speaking realm of Texas.

Contemporary migration to Texas is once again contributing to the linguistic diversity. Immigrants are diffusing a whole host of different languages from their origins around the world. For Texas, the

FIGURE 4-6 Multilingual Signs, Houston
Numerous different languages and scripts in one scene, Harris County. Photo by Author.

introduction of Asian languages such as Mandarin Chinese, Hindi, Korean, Vietnamese, and Filipino are a new phenomenon, which has a strong urban dimension. Emigration from Latin America is not only reinforcing Mexican dialects but also introducing new Central and South American Spanish dialects. In addition, numerous Amerindian languages especially Mayan and Incan dialects as well as Portuguese from Brazil are diffusing into the United States.

English dialects are also evolving in contemporary Texas. The subcultural dialects that defined rural America so well is fading away as new regional and urban realities emerge. The strong distinction between Hill and Plantation South dialects in Texas was already breaking down as many Texans are speaking a combination of the two. Beyond a common Southern experience, the changing realities of economic activities and modern media are propelling linguistic changes. Furthermore, Texans are adopting similar words with all the other America dialects which are converging around a Standard American English (SAE). At the same time, an awareness of Texas identity might allow for some conscience linguistic modifications. Some scholars are suggesting a new Texas dialect is forming with distinctive pronunciations. Known as Texas Vernacular English (TVE), scholars believe it is evolving into a regional norm. Texans seem to be modifying their speech to reflect one's identification with the state, and some of the new distinctions have an urban-rural dynamic. Texas cities are more likely to have SAE elements, but they also have the most cultural and linguistic diversity.

RELIGIONS

Texas' cultural diversity also has a religious dimension. Depending on how one measures diversity, religion can be thought of as one of the more fractured elements of culture because of the long list of church affiliations. On the other hand, much of the organized religion in Texas can be generalized into four major Christian denominations. Generally speaking, Texas reflects an above average religiosity with slightly higher rates of affiliation but not the highest in the nation. Geographers recognize a North–South pattern to religion in Texas as well as some confirmation of national patterns.

Religion is an important element of culture as it contributes to both personal identity and symbolic landscapes. For many, one's religion is the most important component of who they are. Because religion is so deeply felt and often naturalized, it is a delicate topic to discuss. In America where the separation of church and state is used as a shield, most educators just disengage from the topic. On a global scale, it's useful to incorporate religion into the more broad idea of belief systems. Throughout human history, groups of people have had nearly infinite number of belief systems with most being animist and unorganized. Theism and organization are religious innovations with geographical origins and pathways of diffusion. Religion is the most recent and organized structure concerning beliefs, and it is organized religion that dominates the contemporary geography of religion. Religion is defined as the organized system of values and practices involving faith in and worship of the sacred and divine. The "shared system of meaning" in the definition of culture is easily transferred to religion. Religion itself is the system and the shared meanings are the beliefs and rituals of that religion.

Texas's North–South religious pattern derives primarily from the percentage of adherents to the large Christian denominations in Texas counties. While we recognize the denominational differences,

TABLE 4-3 Major Religions of Texas

Religion	Adherents
Baptist	4,500,000
Catholic	4,400,000
Methodist	1,200,000
Pentecostal	763,000
Church of Christ	377,000
Lutheran	302,000
Presbyterian	205,000
Episcopal	178,000
Unaffiliated/Secular	7,900,000

Source: Texas Almanac and Glenmary Research Center.

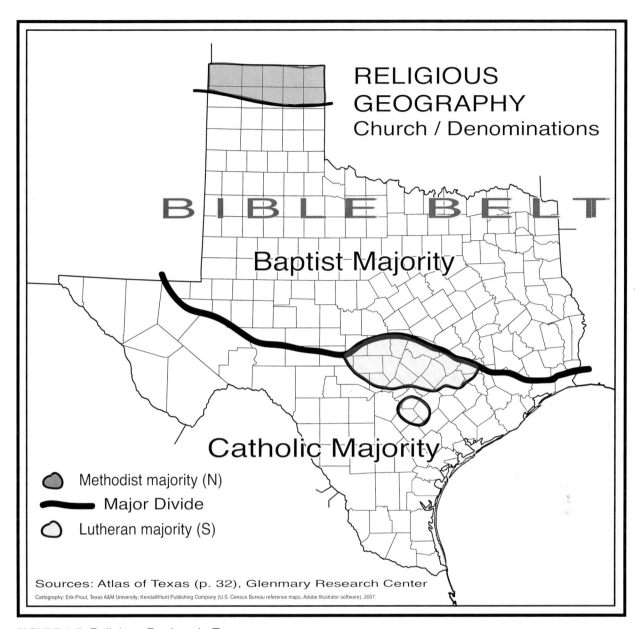

FIGURE 4-7 Religious Regions in Texas

there were also political indicators of this divide. During the contentious votes on Prohibition, the North voted for while the South voted against. The northern half of the state is closely linked to the national vernacular region called the Bible Belt. The Bible Belt spatially covers most of the American South and combines both the Southern experience with American religious revivalism. The strong regional association of Baptists with the South and Bible Belt extends westward into Texas. Baptists are the largest denomination in a majority of counties in the northern half of the state. The next major denomination is Methodist with some counties in the North having Methodist majorities. Often Baptists and Methodists are the second largest denomination when the other is the largest. For the most part, the diffusion of

Baptists, Methodists, Presbyterians, Episcopalians, and numerous other Protestant denominations occurs with the Anglo-American immigration. They reflect the American penchant for high fragmentation and local control over individual congregations. As expected the oldest churches and congregations for these religions are located in the eastern side of the state.

The southern half of the state has two theologically different denominations with dissimilar migration/diffusion stories. The Catholic Church initially diffused to Texas from Mexico during Spanish dominion, which makes it the oldest Christian denomination in Texas. This old heritage is seen visually in the cultural landscapes of missions and churches; one of which is the Alamo. Primarily, Catholicism was associated historically with Mexican-Texans and geographically in South Texas. Secondarily, Catholicism was connected to the Direct European immigrants who emigrated from Catholic areas: Irish, Italian, and some of the Central Europeans. With contemporary migration, Catholicism is on the verge of becoming the largest denomination in Texas—again. The current geographical pattern of Catholicism in Texas and America reflects the growth of the Hispanic population in both rural and urban areas. For American Catholics, the two regional associations are the Northeast (European migration) and Southwest (proximity to Mexico).

FIGURE 4-8 Ysleta Mission, El Paso
Nuestra Senora del Carmen, Corpus Christi de la Ysleta. Photo courtesy of Texas Department of Transportation (Previously published in *Texas Highways*).

FIGURE 4-9 Baptist Church, Independence
Site of Sam Houston's baptism in Washington County. Photo by Author.

FIGURE 4-10 Lutheran Church, William Penn
German settlement in Washington County. Photo by Author.

FIGURE 4-11 Catholic Church, Shafter
Mexican settlement and "Ghost Town" in Presidio County. Photo by Author.

On the other end of the Reformation dichotomy is the Lutheran Church. Named after Martin Luther, who led the Protestant reforms in Germany, Lutheranism took hold in much of Northern Europe. German and Scandinavian immigrants established the Lutheran Church in the United States. Those areas in Texas with Lutheran majorities were typically counties with high numbers of German and/or Direct European migration. Currently, there are only a few counties with Lutherans as the largest denomination. Together, Catholics and Lutherans were the statistical majorities in the southern half of the state. While being very different denominations, they both tended to resist American revivalism and prohibition politics. In effect they became an important check-and-balance to the religious fervor of the Bible Belt in Texas.

Religion inspires symbolic and at times spectacular landscapes. Coinciding with the personal significance, places of worship and sites of important events become sacred. Most individuals follow a reverent code of behavior at these sacred places: for example, one's demeanor in a church and at cemeteries. On occasion, the collective effort of adherents produces grand churches with expensive designs and materials. In many communities, churches are status symbols being one of the tallest or aesthetically beautiful buildings. These buildings contain symbolic elements such as altars, doorways, windows, and orientation to cardinal directions.

Churches in Texas reflect the cultural diversity of the people who constructed them. Rural Protestant churches built by both blacks and whites are simple buildings lacking ostentatious iconography. They are usually white painted wood construction with a steeple, double front doors, and symmetrical, equal number of windows on both sides. On the inside, the pews are lined up in regular patterns all facing forward; up front, they have an open space for quires and a lectern for delivering sermons from.

FIGURE 4-12 Saints Cyril and Methodius Church, Shiner
Catholic Church built by Germans and in cathedral style, Lavaca County Photo by Author

Urban congregations eventually built very big churches with more permanent materials such as brick and mortar. Catholic churches in Texas typically have an outwardly Spanish or Southwest style. The inside contains much iconography, for example, crucifix, stations of the cross, and multiple altars to patron saints. Catholic churches associated with large settlements have a prominent location along the main plaza, which was by design and common throughout Latin America. As part of relocation diffusion, the direct European migrants to Texan-built churches with Old World designs. Similar to their other buildings, they constructed churches with permanency in mind; they also considered prominence with location and height of the buildings. The inside reflects the denomination in terms of iconography and occasionally one sees scriptures or phrases in Latin or German. Outside the church is another realm of religion that will be discussed in the next chapter.

FIGURE 4-13 Rural Protestant Chruch in East Texas
Typical unadorned structure near Burkeville, Newton County. Photo by Author

FIGURE 5-1 Big Texan Restaurant in Amarillo
Texas stereotypes are numerous at the Big Texan along busy IH-40, Potter County. Photo by Author.

5
Cultural Landscapes

Texas has an interesting variety of cultural landscapes. This variety has its origins in both cultural diversity and differences in physical environments. One can still observe the different folk cultures across rural Texas. However, there are great changes to everyday landscapes, and older vernacular landscapes are being lost to modern, standardized ones. Chapter Five surveys some of the distinct landscapes in Texas. The approach taken in this chapter is to discuss five specific topics that illuminate the cultural geography of the state. First, the idea of landscape will be defined and the overlap with settlement geography will be clarified. Then, we will explore cemeteries, survey systems, folk architecture, courthouses, and ranching in that order.

CULTURAL LANDSCAPES

Many geographers associate the study of cultural landscapes with Carl Sauer and the Berkeley school of geography. Carl Sauer was a professor at the University of California, Berkeley who became a major figure in American geography. Sauer had numerous students who went on to influence the course of many different geography departments around the country. While cultural geography, cultural ecology, historical geography and even foreign area dissertations are indebted to Sauer, it all comes back to landscape. The cultural landscape is more than a topic; it is interwoven with the methodology of observing people and places. The essential idea is that the landscapes or the immediate environments around humans reflect their cultural uniqueness and interactions with nature. The cultural landscape is a composition of new and old human modifications to the Earth, which has meaning to those who produce, contest, and consume them. The cultural landscape is foremost a lived in and experiential thing that researchers can exam, record, describe, explain, and ultimately understand.

$$\text{Human Culture} + \text{Natural Landscapes} = \text{Cultural Landscapes}$$
$$\text{"Agency"} \quad\quad \text{"Medium"} \quad\quad \text{"Result"}$$

Sometimes it is useful to have landscape ideas elaborated. In his well noted work, Sauer outlined cultural landscapes as the product of human agency and natural media. This can be presented like a mathematical formula, even though it's not meant to be quantitative. Human agency was argued to be culture, and culture is a part of our decision making about what to do (actions). Natural landscapes are the underlying media that humans interact with as we live on this planet. Therefore, the existing cultural landscapes that we see are the result of agency and medium—both in the present and in the past.

When we look at a cultural landscape, sometimes we have to interpret multiple layers of meaning. At certain places, the layers can be literal time periods when different influences shaped the landscape. For example, a place such as Mexico City has an Aztec layer, a Spanish colonial layer, and then an independent Mexican layer. Those places in Texas old enough might have indigenous, Spanish, Mexican, and American layers. Geographer, Derwent Whittlesey coined a term for this: *Sequence Occupancy*. The landscape one looks at is being shaped by people in present, but the previous occupants also had a role in shaping it. The further in the past sequentially those influences go, the more masked or hidden they are due to more recent occupants. The photo of the Alamo provides a discussion point for Whittlesey's sequence occupancy. Originally constructed as a Mission, the Spanish architecture of the chapel façade is the most iconic element of this landscape. With the Daughters of the Revolution maintaining the main building as a shrine, the surrounding grounds and preserved features are all the product of this group. Framing the photo is the reality of a major urban area that grew around the site.

Settlement geography is the study of how people create their own habitats as well as the corresponding patterns and impacts. The rationale is that humans arrange themselves on the Earth's surface, and there are some basic geographies to it. One is that people congregate together into social units such as family and community; these units have corresponding settlement features such as farmsteads and towns. As one could anticipate, rural and urban settlements would differ in ways beyond just population density. Agriculture and other resource extraction activities tend to dominate the nature of rural settlements; meanwhile, the scale of

FIGURE 5-2 Alamo Plaza, San Antonio
The Alamo and its current urban setting; public plaza being used for a Christmas tree, Bexar County. Photo by Author.

production and consumption impacts urban settlements. A second geography refers to the explanation of the patterns. Settlement geography strongly reflects the people or agency of culture, reflects the natural environment, and finally reflects the activities that they are engaged in. A third aspect of settlement in Texas is time. Compared to other places on the planet, settlement in Texas is very recent—only in the scale of hundreds of years. However, the frontier nature of the settlement meant that many of the earliest structures and places were not meant to survive. Settlement geography by its very essence as human presence is observable, which means the techniques of landscape analysis are transferable. In effect, the cultural landscape is the researcher's evidence for settlement. We now turn to the first of five landscape examples that also illuminates the settlement geography of Texas.

CEMETERIES

In the previous chapter, the geography of religion in Texas reflected the distribution of Christian denominations. Across the state, special places associated with congregation are one of the more symbolic buildings in Texas, and most people consider them to be sacred. Sacred places have their own code of conduct that differs from profane places. Closely associated with religion are the sacred beliefs and practices that we have for the dead. In fact we have multiple geographies of the dead—sometimes collectively called *necrogeography*, which include the literal geography of mortality. Another such necrogeography is the afterlife or spiritual destination; for example, many Christians have a paradisiacal image of heaven with pearly gate borders and inhabited with peaceful souls. Of course there's the other afterlife destination with opposite, unpleasant images. Other religions have afterlife images as well, but most all have another worldly dimension to them. With bodily death, the survivors have to deal with the actual flesh and bone bodies. A feature of human culture is our recognition of death, and a deep motive to remember the dead. One aspect of respecting death is to treat the dead body with respect. The initial memorialization process

FIGURE 5-3 Southern Folk Cemetery, Independence
Entrance gates to Independence Cemetery, Washington County. Photo by Author.

FIGURE 5-4 Southern Folk Cemetery, Steep Hollow
West facing tombstones in Steep Hollow Cemetery, Brazos County. Photo by Author.

often begins in a place of worship, once again linking death to religion. The next step in the process is a burial ceremony usually led by a religious figure. This leads us to another necrogeographic feature that have a strong landscape dimension—cemeteries.

Cemeteries or graveyards in Texas have their own geographies. At the most basic scale, one can ask where are the bodies? Cemeteries differ in the arrangement of bodies. Cemeteries differ in the types and arrangements of gravestones. Cemeteries differ in their proximity and relationship to a church or community. We can account for much of the cemetery differences by generalizing them into three culture group categories. However, the modern cemetery is becoming a major if not predominant type. Therefore, we will discuss four cemetery types: Southern folk, Mexican/Catholic, German, and Modern.

Southern Folk Cemetery

The Southern folk cemetery is the most frequent cemetery type in Texas, and it can be found all over the state. The cultural diffusion of most aspects occurs with the migration of Southerners into Texas. It is a folk cultural phenomenon with everyday people who weren't consciously designing cemeteries. In addition, numerous families created their own burial plots which may have evolved into community and/or congregational sites. Therefore, there is much variation between cemeteries as different communities dealt with local terrain and cultural diversity. The geographical concentration of Southern folk cemeteries coincides with the Bible Belt and its large rural Baptist population. As with other things, segregation between blacks and whites is very notable with different cemeteries or separate sections of cemeteries. This is quite ironic because they both have some common features.

Southern folk cemeteries have a host of features but they are not universal to each and every cemetery. One of the outstanding features is the alignment of bodies with the feet pointing eastward. The explanation for this alignment is that one will be facing East toward the Lord on judgment day. Another feature is the direction of the writing on the grave markers, which in Texas is often on the head side of the body. The larger cemetery design includes a subdividing of the land into family sections that are often demarcated with fences or curbs. Cemeteries also have a temporal dimension with the older individual and family graves in one part of the grounds and more recent plots with newer tombstones in another. A ceremonial main gate is a mainstay with its ironwork arch over the entrance; the rest of the fence is usually a practical material to keep animals out. Many of the design features of southern folk cemeteries blend into generic rural cemeteries of other cultural groups such as main gates and gravestone writing. On the opposite end, scraping and cleaning days didn't diffuse and are being lost. Scraping is the removal of all vegetation around the gravesite, and often survivors place objects such as sea shells to serve as both decoration and deterrent to vegetation. Scraping was practiced by both blacks and whites across the South, but its origin was most likely with the Afro-Caribbean slaves. Cleaning days are community designated dates to collectively clean up the grave markers and cemetery grounds. Cleaning days are most likely to be found in Northeast Texas.

Mexican and German Cemeteries

Two other folk cemetery types in Texas are associated with large regional concentrations of migrants from Mexico and Europe. Mexican migration produces the dominant Catholic cemetery type in Texas as well as

FIGURE 5-5 Mexican Cemetery, Socorro
Socorro Mission, El Paso County. Photo by Author.

FIGURE 5-6 German Headstone, Serbin
Shared headstone for three siblings in 1883, Lee County. Photo by Author.

the oldest gravesites in the state. As the largest direct European group, we call the third folk cemetery German, but there are commonalities to the other European groups. All of these cemeteries are likely sites to find non-English headstones, which might have a special research aspect of not being lost in translation.

Mexican cemeteries are linked to the Catholic church's organization such as a diocese or parish. The land is considered sacred and usually goes through a sanctification process. Typically, the cemetery is located adjacent to the church or mission, but in other instances, there are rural sites and segregated sections of community cemeteries. Sometimes Catholic cemeteries have family plots, whereas others follow a temporal order. It is common to have a "little angel" section where children are buried. Mexican cemeteries in Texas appear to be more lavishly decorated than others. Fresh flowers, candles, and decorations are found on both recent and not so recent gravesites. Like in Mexico, there are numerous activities around All Saints Day and All Souls Day, which usually includes a visit to clean and decorate close relative's gravesites.

German cemeteries in Texas stand out for their orderliness. Like European cemeteries, they are very structured with efficient use of space. The sight of burial plots being lined up in straight lines is very apparent. Moreover, there is a strict temporal order to the burials with individual sites organized by the calendar date of death. This is useful when trying to estimate tragic events and epidemics. Despite the generalized rigidity, German tombstones show individuality in terms of tombstone shape, size, and decorations. Over time, German cemeteries allowed family plots and other southern features.

Modern Cemeteries

Modern urban Texas presents us with a different dynamic than rural communities had in the past. It is true that rural communities had to deal with the reality of cultural diversity when it comes to cemeteries because they had to respect the dead and preserve health standards for all. In effect they provided a model for the future. They created community cemeteries that allowed burials from different congregations, yet they often were segregated by ethnicity and race. Urban areas with millions of citizens poses an even more daunting task of creating and maintaining large enough spaces for the number of dead. The usual death rate is one percent of the population every year, which means for a metropolitan area of 5 million approximately 50,000 burial sites.

The emergence of private cemeteries is one dimension to this large task. The most obvious geography is the scale or size of modern cemeteries—both public and private. The landscape is typically more open and organized into large sections. It is common to have an onsite chapel and paved roads. One strong

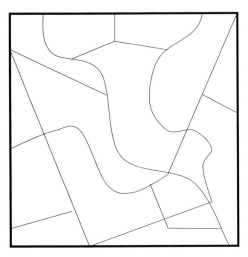

Metes and Bounds
Initial Anglo-Southern influence in early settled areas (formal boundaries are often recorded after settlement)

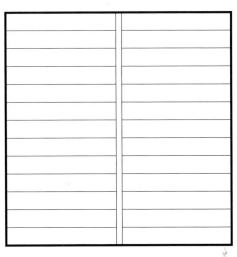

Long Lots
Spanish and French influence along waterways and historical roads (typically surveyed prior to settlement)

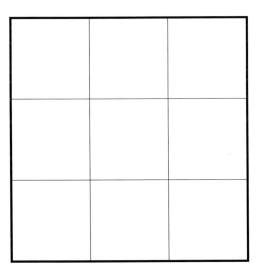

Regular Rectangles
Modern American survey system that attempts to divide land equally; found in western half of the state (surveyed prior to settlement)

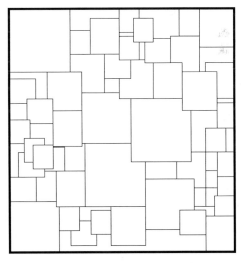

Irregular Rectangles
Anglo-American influence; found in eastern half of the state (reflects intense subdividing of land after settlement)

FIGURE 5-7 Land Survey Systems in Texas

trend is toward horizontal grave markers, which have the advantage of easier landscape maintenance. Except for recent burials, no decorations or objects that would block lawn mowers are permitted. Many large cemeteries have mausoleums for the interment of ashes and small memorial markers. Undoubtedly, burial practices will continue to change, and urban cemeteries will probably lead the modern trends.

LAND AND ORGANIZATION

Is there an underlying order to the cultural landscape in Texas? Yes, humans structure their activities on the Earth's surface. Perhaps the key idea that influences landscapes is private property. There is a relatively recent historical dimension to private property because it wasn't present 500 years ago and it diffused to Texas with European and American influence. Technically, private property includes both land and personal possessions, which are connected to other major ideas like privacy, resource, and homestead rights, but the focus here is on the land property. It is commonly understood that the land can be designed and demarcated for optimal ownership and use. Two examples of how this rational thinking shapes the landscape are survey systems and city planning.

Surveying includes the actual techniques for measuring land, demarcating property lines, and translating that into the legal written and graphic descriptions of property. A survey system exists when a predominant surveying method is applied over large areas, which is the case in Texas. As the state sold and gave away land, they were able to survey the land prior to settlement that resulted in consistent shapes and sizes. However, other survey systems were already being used. Texas has examples of four different survey systems each with their own geography.

Survey Systems

The first survey system is known as Metes and Bounds, which is a continuation of oral culture societies. If two settlers agree that the border between their properties begins at the big oak tree on up to the top of the hill, then that is Metes and Bounds. Property lines with this system are not systematic and they may even not be straight lines because they follow natural features such as creeks and ridgelines. Afterwards, the need to survey the property line for official records puts the oral description into a legalistic terminology. Metes and Bounds system is located in the older settled areas of Texas—both Mexican and Anglo. It is more prevalent in East Texas along roads such as OSR that the initial Southerners used to enter Texas.

The second survey system is known as Irregular Rectangles. The idea is to create square shaped properties that can be easily surveyed and still have the flexibility to own different sized lots. The surveying of the property is done before most of the settlement has occurred, but the system allows for the subdividing and consolidating of property into the current irregular property line patchwork. Geographically, Irregular Rectangles are found on the eastern side of the state, which corresponds with the Anglo-American settlement on the coastal plain.

Long Lots is the third survey system found in Texas. Geographers associate Long Lots with French and Spanish colonies in North America. The Spanish introduced them into Texas first, but migrants from Louisiana also contributed to the design of coastal plantations. Long Lots are also rectangular shaped but the distinctive element is the length/width ratio: typically a 10 to 1 ratio. Long Lots are often found along a linear transportation feature such as a road or river. Each property owner has access to the road or river for transporting goods to market while maintaining a large enough property to be viable. Long Lots in Texas are found along the earliest settled river basins; Rio Grande, San Antonio, Guadalupe, Colorado, and Brazos Rivers.

The final survey system is Regular Rectangles. Like the previous two, Regular Rectangles are geometric, but they are meant to be consistently equal in size and nearly square in shape. Regular Rectangles

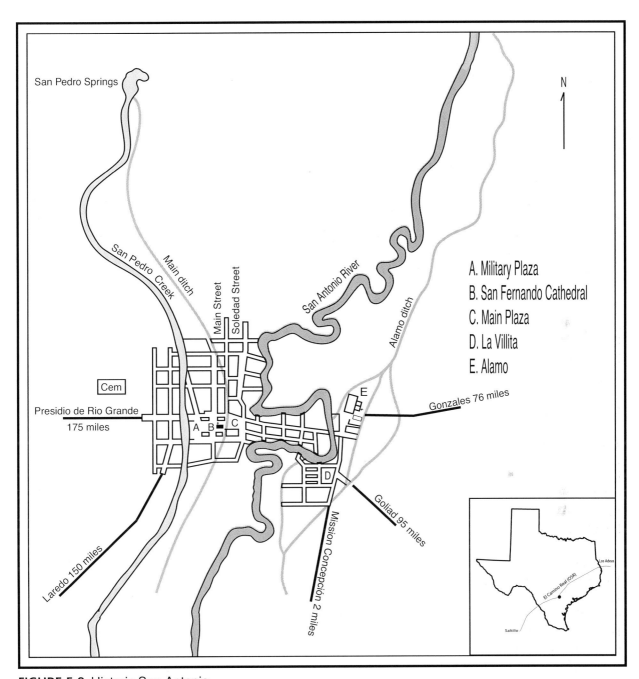

FIGURE 5-8 Historic San Antonio
Source: *Historical Atlas of Texas* (map 13). San Antonio consisted of various Spanish institutions, including multiple missions, a presidio, and a chartered town (La Villita, 1731). Original layout of city and plazas along San Antonio River as well as Alamo ditch and San Pedro Creek is apparent. The relationship to water was obvious, and the layout of a street grid with plazas conformed to Spanish policy.

are found in the western half of the state, and they are the result of government control over the surveying and settlement. For example, much of the Panhandle was organized with one legislative act that included both property and the counties. Regular Rectangles produce the consistent agricultural patchwork pattern found throughout the middle of the United States.

FIGURE 5-9 Main Plaza, San Antonio
Bexar Country Courthouse is in the foreground; main Cathedral is to the right. Photo by Author.

City Plans

Surveying also applies to property in towns and cities. Property lines in cities often begin as carefully laid out plans, and then they expand while taking into account population growth, economic change, and topography. Urban areas are characteristically more densely populated with a variety of landuses in close proximity to one another than rural areas, which compel government to regulate them better. Prior planning is the best way to achieve this goal.

Spanish colonial authorities drew on both classical city ideas and practical experience of conquest. During the *Reconquesta* of the Iberian Peninsula, Spanish authorities understood the importance of using cities as centers of control just as the Romans did. After their initial contacts in the Americas, they codified a set of principals for all cities with the Law of the Indies in 1573. Once they embarked on large scale colonization, this basic city plan diffused with them. Often called the Spanish city model, it became the most common design in all of the Americas. The plan set out a simple grid pattern with streets intersecting each other at 90-degree angles. In the center of the grid was an open block that would be the plaza. On one side of the plaza would sit the parish church that creates one of the most common landscapes in the Western Hemisphere. The other three sides of the plaza were very desirable locations for government buildings and elite housing. The oldest government building in the United States is the governor's office/residence along the main plaza in Santa Fe, New Mexico.

The Spanish city model diffused to Texas during the Spanish and Mexican periods and numerous places reflect that fact. Spanish Missions did not necessarily follow the official city models because they were autonomous units. Most missions were aligned to rivers and irrigated agriculture; the spatial patterns reflected a social order of the missionaries. However, as communities grew around missions, the grid pattern of streets and central plazas appear. The photograph of San Elizario shows the plaza anchored by the church, which currently contains a public gazebo/bandstand surrounded by pathways lined by with benches. The initial plan for Laredo shows the Spanish city model in an idealized form. As the centers of

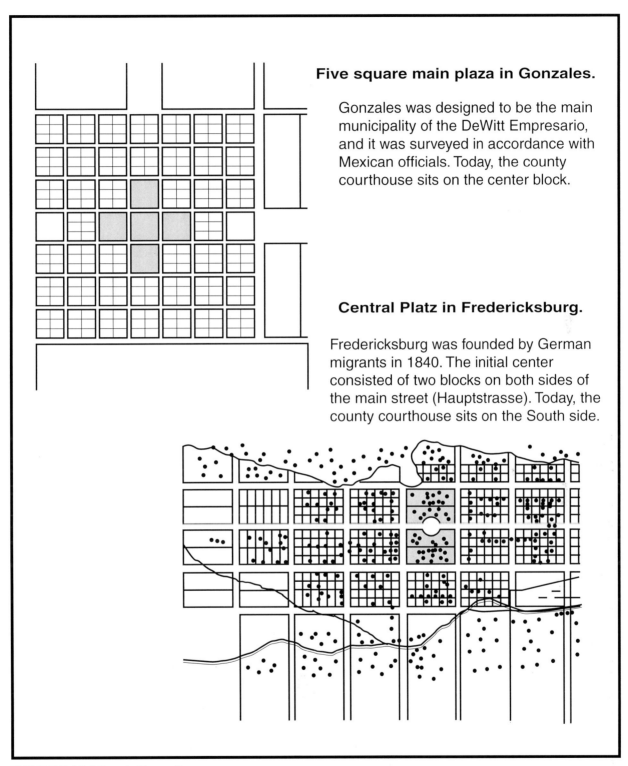

FIGURE 5-10 Historic Town Squares
Source: R. Veselka, *The Courthouse Square in Texas,* 2000 (p. 132 and p. 138). Mexican- and German-influenced courthouse squares are exemplified by the Gonzales and Fredericksburg city plans.

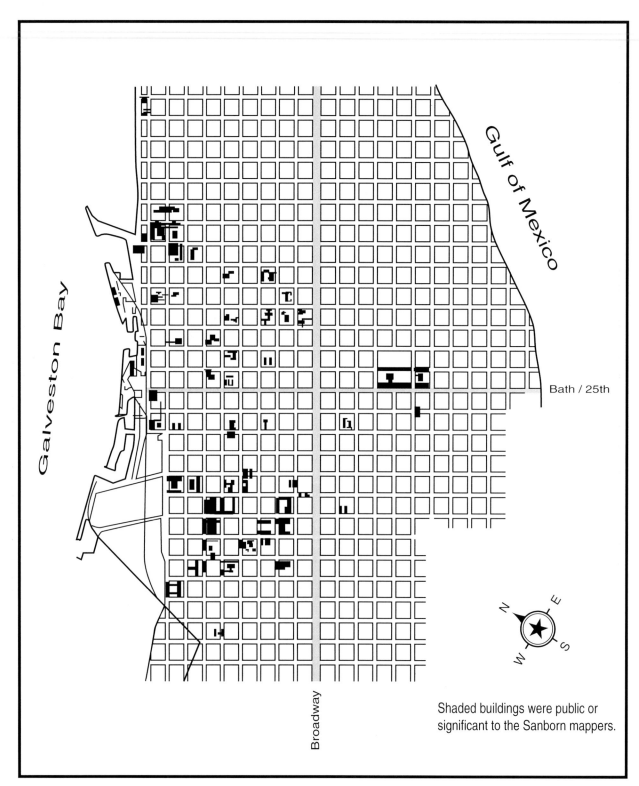

FIGURE 5-11 Galveston City Plan
Source: Sanborn Map and Publishing Company, 1885. These fire insurance maps accurately depicted the building environment as well as the hazards and occupants. Galveston was centered on the bayside of the island, with the merchants and ship docks as focal points.

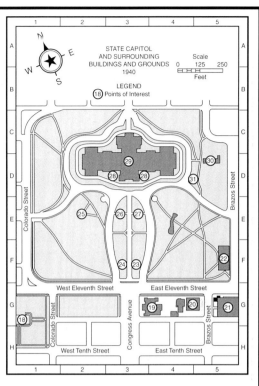
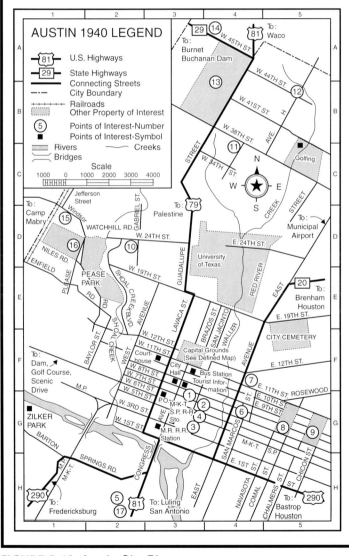

Austin was selected as the Republic's capital in 1839. The initial grid of streets was platted along the Colorado River. Congress Avenue linked the river with the designed Capitol.

Selected Sites:

8 Texas State Cemetery
13 Austin State Hospital
18 Governor's Mansion
22 Old Land Office Building
29 Texas State Capitol

FIGURE 5-12 Austin City Plan
Source: Texas Writer's Program (WPA) Main map shows the inner city before freeways; the layout of major roads and initial grid pattern along the Colorado River is apparent. Inset map is the designed Capitol Grounds with the adjacent state government infrastructure

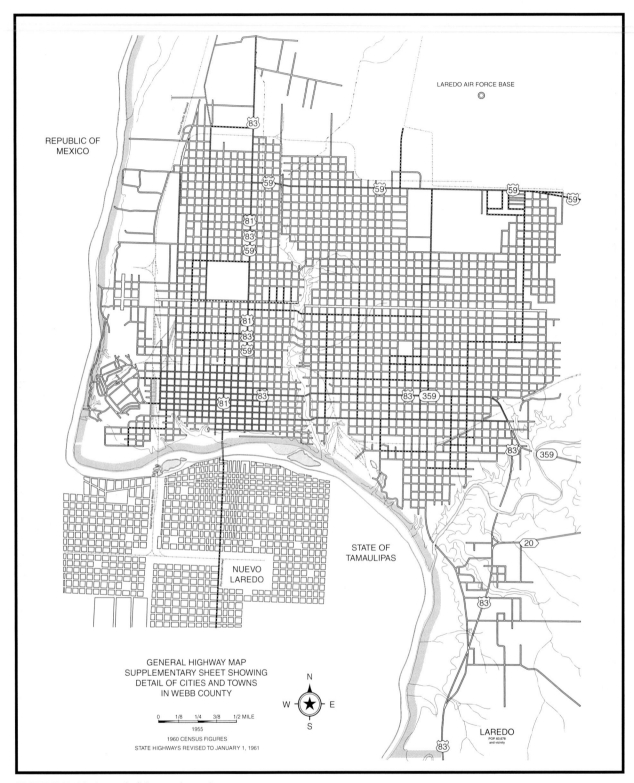

FIGURE 5-13 Laredo/Nuevo Laredo
Source: Texas State Highway Department, 1961 The growth of the city beyond the Spanish plaza core is noticeable. Nuevo Laredo is located immediately across the river, and the two cities really function as a single system.

the DeWitt and DeLeon Empresarios respectively, Gonzales and Victoria were designed with central plazas. Gonzales' plaza actually had five squares in a cross shape pattern, and later the county courthouse was constructed in the middle square. The plaza in Victoria still exists but has the look of an urban park with much vegetation. San Antonio is a more complex example because it developed multiple plazas; the grid patterned central areas had to deviate around the river and Alamo.

The grid pattern was not unfamiliar to Anglo-Americans arriving in Texas. Rectangular survey systems were commonly applied to towns and cities in the South. The sizes of the rectangles were modified smaller to accommodate individual ownership and population density. Role model Southern cities such as Charleston and New Orleans had grid patterns in their historical centers. Perhaps the biggest difference with the Spanish model is that the Americans placed a secular symbol in the center. Those places that would become county seats have the county courthouse located in the center of the grid. Churches were rarely placed in the center of the grid and/or adjacent to government buildings. Economic activities typically surround the central block with one of the streets commonly named Main Street where many business and market activities congregated.

One of the big influences in city form in Texas was the railroad. After initial settlement that was focused on rivers and roads, the railroad became a crucial factor in settlement patterns. First, the actual location of the railroad set into play those places that were connected and more likely to grow. Second, the railroad itself acquired land and could develop places. Therefore, the railroad becomes an agent for diffusing city plans. Railroad induced plans commonly had grid patterned streets that would be aligned to the railroad. Two different generalizations exist in Texas. One is a parallel running Main Street with linear interactions such as warehouses or cotton gins along the rail line. If a public space existed, a plaza or square would be adjacent to the passenger depot, and this is typically the central market space. A different way to plan a railroad town is with a major street to be perpendicular to the rail line. Known as a T-town in much of the United States, the perpendicular street is usually named Main Street (i.e., Disneyland).

FOLK ARCHITECTURE

The earliest Texans constructed very distinct landscapes from one another that reflected their respective cultures. It wasn't because they spoke different languages; it was because they made different artifacts. They built houses, barns, and fences that looked different, and they built them differently. Furthermore, they weren't using blueprints and supply stores. Individuals were constructing artifacts in ways they were familiar with, which limits them to what they previously experienced. Folk architecture includes the designs, the construction methods, and the resulting structures that are produced by folk cultures. As a way of remembering, it is architecture before there were architects. As the different culture groups migrated to Texas, they diffused their folk knowledge about how to produce dwellings and structures as well as their preferences of how they should look and for what function.

The cultural diffusion of folk architecture provides us with a strong evidentiary argument for differences that are hard to distinguish now. Most people live in homes that are nearly identical to ones that other Texans live in despite the unlimited superficial elements such as color. The current distinctions tend to be socio-economic ones that have class connotations: McMansion, trailer-trash, etc. The presence of folk architectural artifacts documents and differentiates the movement of Upper and Lower Southerners as well as the Central European and Hispanic influences in Texas. However, the existence of these artifacts is declining because they are not in use currently and they were built with materials such as wood that decay.

Both log cabins and wood fences fall into this category. If still present, these artifacts can be interpreted for cultural origins. Unfortunately they have decayed if left in disrepair or they were dismantled as newer structures were built. Despite being difficult to find, log cabins are instructional to us because they highlight two elements of cultural diffusion. Firstly, there is an environmental dimension to certain artifacts. In the case of log cabins, the availability of wood becomes a large factor in their presence and size. The forests of East Texas were ideal for log cabins and the Post Oak/Cross Timbers were suitable, but the lack of forest elsewhere necessitates other forms of shelter. Secondly, the technical details of log cabin construction can provide researchers with nuanced specifics. Stacking logs is inherently dangerous and it requires a proper notch to stabilize the structure. This notch interlocks the perpendicular logs at the corner of the structure. It turns out the notch is a very culturally specific artifact that can used to distinguish between people who may not be aware. For example, the log cabin is strongly associated with the Mid-Atlantic subculture that is very amalgamated by the time Upper Southerners arrived in Texas; the notches differ between Scots-Irish, Germans, Finns, etc.

Log cabins were initially used as primary dwellings for nuclear families, and over time, they could be enlarged. By the end of the 19th century, there was a stigma associated with living in log cabins, so most families built frame construction houses as soon as they could. Nevertheless, rural poor continued to use log cabins through the Great Depression. Eventually, older cabins were modified or converted to barns, storage, etc. Like log cabins, fences had an ephemeral fate. Fences had a more circumscribed role than they do now. The earliest fences were meant to keep animals out—not in. Domestic animals were let loose to fend for themselves (take root) and the fences were only to keep them out of the vegetable gardens. Occasionally, it was necessary to round up and brand the animals for ownership purposes, which was the purpose of Southern cowboys. Eventually pens were constructed out of wood as well. After alternative materials such as barb wire became feasible, people began to fence in their animals and fences coincided with property lines. Wood fence types were culturally specific and for Texas had a strong connection to the Upland South. We could probably go so far as to say forest resource use and skilled carpentry diffused as a

FIGURE 5-14 Dogtrot Cabin at L. B. Johnson Ranch
Lyndon B. Johnson State Historic Site, Blanco County. Photo by Author.

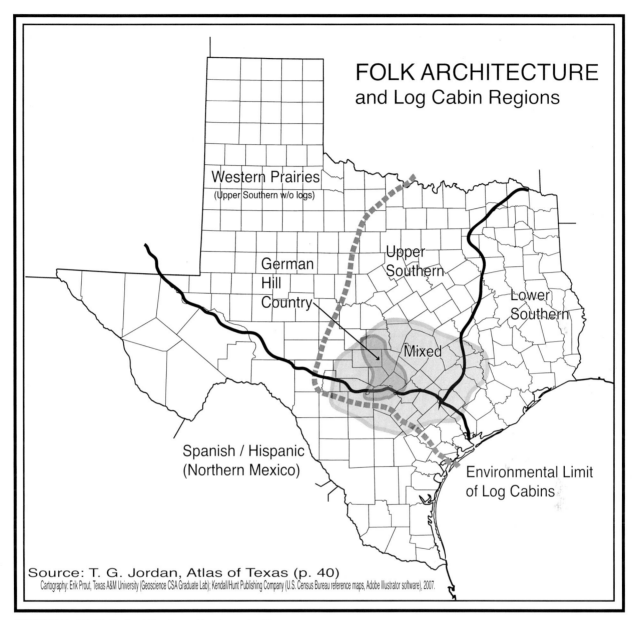

FIGURE 5-15 Folk Architecture Regions in Texas

complex with Upper Southerners. However, both Upper and Lower Southerners as well as Europeans built log cabins. Texas is the product of multiple migration movements and they all had their own contributions.

Taking into account the various groups and artifacts, there are four folk architecture realms to Texas. Mid-Atlantic folk architecture diffuses to Texas primarily with Upper Southerners and secondarily with Lower Midwesterners. The previous examples of log cabins and wood fences are associated strongly with this group especially those originating in Appalachia. Floor plans such as the I-house are as well. Blacks and whites from the Lower South introduced many features that became trademarks of Southern houses. The extension of rooflines to create a patio or porch space that is protected from the elements is one such feature. Another feature is the use of pier and beam construction that lifts the floor

FIGURE 5-16 Main Street (Hauptstrasse), Fredericksburg
Continuity of the built environment reveals German folk architecture; Gillespie County. Photo by Author.

up off the ground surface. A tendency to separate the kitchen from the main house is a third feature, which had a social function. These features can be seen on the dwellings that range from the elegant plantation home to the simple shotgun house. Also from the Lower South, many of the initial urban forms for symbolic buildings, central business districts, and residential segregation diffuse to Texas.

The architectural traditions from both Hispanic and Central European migrants include an urban dimension. Perhaps we see their architectural contribution in specific regions of rural Texas better, but they did influence urban Texas. At various times, German migrants were the largest culture group in both Galveston and San Antonio. Germans, other Central European, and Americans from East coast cities settled in Galveston, which became the most urbanized city in Texas, and then some of those urban ideas diffused to budding towns and cities throughout the state. In the case of Germans, scholars can distinguish a folk architecture realm that correlates with Hill Country. In this realm, both a rural and urban dimension stands out. The initial European settlers built log cabins and they had distinct notches from their Anglo-American contemporaries. Those who specialize in these artifacts can even distinguish between the Medina (Bandera and Castroville) and Guadalupe River settlements. In places such as Fredericksburg, the brick and stone multistory buildings stand out for the position and permanence. In general, Germans constructed with more permanent materials and they were more likely to build second floors at the time of initial construction. *Fachwerk* or half-timbering is one of the signature building techniques.

Spanish style architecture of Missions was clearly the first European influence to diffuse into Texas. Moreover, there is a Hispanic folk architectural realm that covers South Texas and along the Mexican border. In fact much of northern Mexico and the American Southwest are climatically similar to the Iberian Peninsula; traditional Spanish land uses and construction techniques are well suited there. Bricks and blocks made of clay or mud are the primary building blocks. Then, stucco or plaster is applied as exterior surface to walls. Generally, the lack of forest means that long and thick wood beams are rare, which

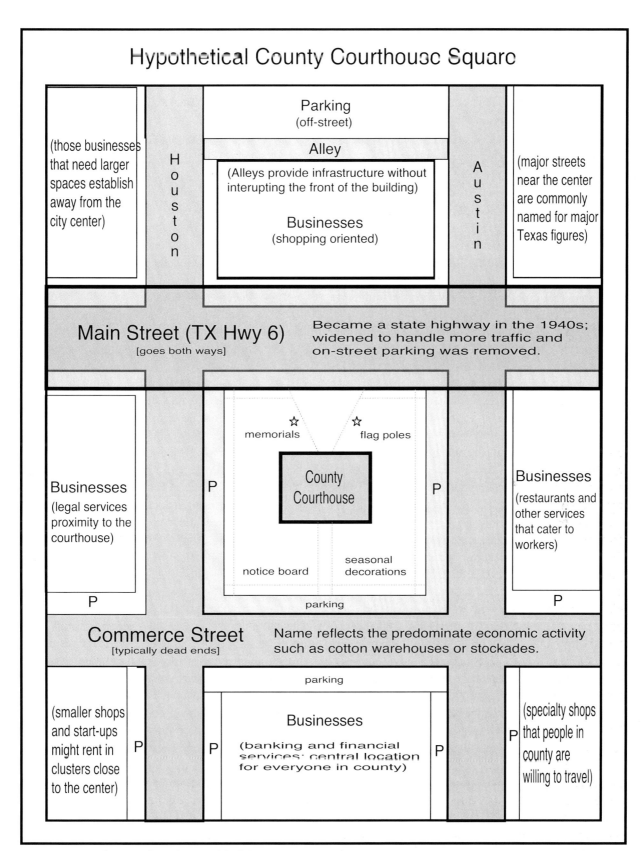

FIGURE 5-17 County Courthouse Model

makes roofs distinguishing. Flat roofs are more common in the interior or far west Texas than South Texas. However, smaller pieces of wood from Mesquite for example could be tacked vertically to make fences or integrated into building walls; this is known as palisades. Ceremonial buildings such as the church or governor's residence were more likely to follow designs from Spain and/or Mexico even if they had to be built with locally acquired materials. Later when county courthouses were being constructed building materials could be transported in.

COURTHOUSE SQUARES

Courthouses are significant buildings in their own right because they literally house the judicial branch of our government. Throughout America, courthouses are typically imposing buildings with symbolic architecture with classical elements making reference to Greece and Rome. Many have iconic artwork and quotes related to liberty, justice, and democracy. County courthouses became synonymous with local government in Texas. The lack of local government was one of the biggest grievances that Anglo-American settlers had with Mexico; during the Republic only two levels of government existed: Congress and counties. The creation of new counties and the selection of county seats was a constant political issue in the 19th century. While we focus on county courthouses (CCH), there are state and federal courthouses with their own symbolism; it is at the county scale that we see community. Ironically, it is the State Capitol that serves as a larger scale model for Texas courthouse landscapes because of the building's grandeur and centrality to Austin as well as the controversies about the monuments in the public spaces.

County courthouses in Texas are typically centered in the middle of a city block that is usually just one of the blocks created by the grid pattern of streets. However, there are other prominent ways to situate the courthouse in relation to the street grid. Some of the more common types are shown in Figure 5-17. Courthouse square types are named after the oldest known example, which informs us that the Mid-Atlantic subculture diffused most of them to Texas. The Shelbyville (Tennessee) is the most common type in Texas, which conforms to the CCH model (Figure 5-16). The Lancaster and Harrisburg (Pennsylvania) are two types that align the courthouse with a main street to create a visually dominant courthouse along the streetscape. Multi-block (2–6) and partial-block (1/2–1/4) courthouse squares exist as well as integration with Spanish plazas.

Of course the buildings themselves are often the visual focus of the squares. Much of the scholarly interest revolves around the historical and architectural aspects of the buildings. Almost all courthouses in Texas are not the original ones—commonly the second or third iteration. Most communities rebuilt their courthouses as a statement of civic pride and they hired renowned architects to design newer and bigger buildings. To some degree these new courthouses reflected a vision of progressive urban realities. Now that these buildings are relatively old, recent preservation efforts are propelled by a conservative rurality. The courthouse has come to symbolize all that is right and stable in the face of constant change that modernization brings. Despite this complex motivation, historical preservation is concerned exclusively with the structure.

Many county courthouses are surrounded by landscaped grounds that contain numerous icons. The grounds are well kept up and even the focus for civic groups to beautify with flowers. Memorials ranging from Civil War statues to founders plaques are very common, and many counties have multiple official state historical markers. A definite trend around the state is the erection of a war memorial to commemorate veterans, military branches, and war dead from the county. The grounds are still used by the community for ceremonies, public gatherings, and public displays such as Christmas scenes.

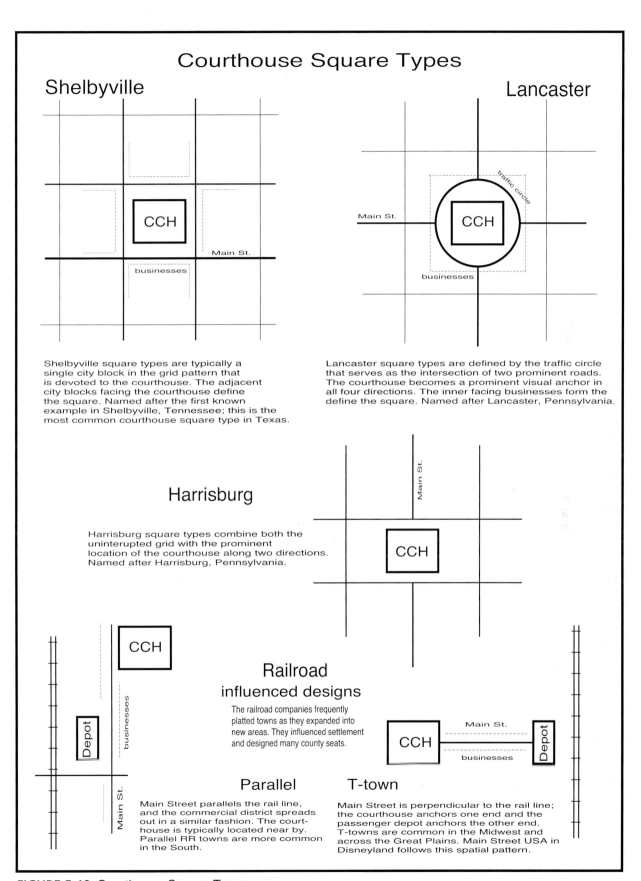

FIGURE 5-18 Courthouse Square Types

FIGURE 5-19 Christmas Decorations in Courthouse Square, Uvalde
Public space around Uvalde County Courthouse is used for seasonally themed decorations and community events. Photo by Author.

FIGURE 5-20 Presidio County Courthouse, Marfa
Recent preservation efforts have a noticeable effect on courthouses around Texas. Photo by Author.

Immediately adjacent to the courthouse grounds is the street grid, and in the case of the Shelbyville model, four blocks adjoin forming a square that faces the courthouse. These four sides become the primary commercial and business zone for many county seats. In addition to law offices and government annexes, a bank, hardware store, restaurant, and small shops are found here. In some counties, the square is extremely vibrant and remains the center of community activity. In other counties, commercial activity has moved to a more accessible part of town along the highway. In these instances, the square has only minimal activities such as antique stores and even residential homes.

All together the complete courthouse square includes the courthouse building, the actual block, and the adjacent public square. With preservation efforts to the buildings, the courthouse is likely to remain a significant office space for local government. In many county courthouses, especially smaller places, other government activities are already housed in the building. A few counties transform the courthouse grounds and square into a broader government center. One such theme is to build community oriented things such as a library, swimming pool, boys and girls club and senior center on the actual grounds or adjacent property. Another common theme is to enhance the law enforcement function with expanded jail operations and police stations. However, many counties create a modern law enforcement complex away from the courthouse which allows for other governmental and ceremonial activities at the courthouse. As

FIGURE 5-21 Ranching Scene
Cattle and fence lines along US-183/77A in Goliad County. Photo by Author.

the county courthouse remains the predominant cultural icon for Texas, the wide open spaces associated with ranching influences our images of nature.

RANCHING

One of the most stereotypical images of Texas is the cowboy. It is the cowboy that provides a human face to ranching, rodeos, and the open-range landscape of cattle drives. Of course the cowboy has a gendered counterpart in the cowgirl. The contemporary images of cowboys and cowgirls mask the realities of harsh

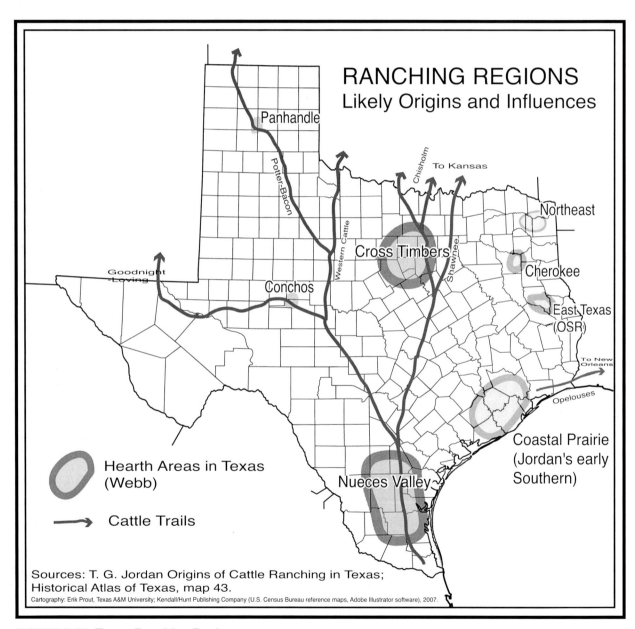

FIGURE 5-22 Texas Ranching Regions

social relations that produced ranch hands as a type of paid and unpaid labor. Still there is enough skill and tradition in the practices for many to take pride in the lifestyle. In this section, we discuss the origins of ranching in Texas because it reinforces the dominant themes of this chapter. Once again we have another cultural element in Texas that is complex with multiple diffusion pathways.

In many ways, ranching is the most ubiquitous cultural landscape in Texas. It is a cultural landscape because human agency in the form of land management, property ownership, and animal husbandry shapes the appearance. The rancher controls the number and type of animals through buying, selling, and reproduction; the rancher also places the animals in defined spaces enclosed by fences. Ranchers have modified physical systems with water capture and vegetation composition for example. Ranching is ubiquitous or very common sight around the state because it occupies more land than any other activity in Texas. Texas overwhelmingly has more acreage in agriculture than any other U.S. state, and furthermore, the Census classifies over half of the land area as rangeland. The extensive and expansive use of land for ranching is well suited in much of Texas. Those places with good soil quality and moisture availability will have other agricultural activities, but the value of crops is less than half of livestock value in Texas.

Without a doubt, ranching is significant to both the economy and ecology of Texas. It is also a stereotyped image of the state that reinforces the notions of authenticity toward identities and practices. Many people associate Texas with ranching, and it clearly has played a leading role in the development of ranching in the United States. However, ranching did not originate in Texas. The question arises from where and when did ranching diffuse to Texas? Then, one might ask how did it evolve and eventually diffuse throughout the western United States? It turns out there are two strong inputs that diffuse to Texas.

The first component is the Spanish herding tradition. In Spain, cattle, sheep, and goats were the primary herding animals; other domesticated animals included pigs and chickens. The Shepard tradition was the dominant way to tend animals, and many places communally organized their grazing and milking. It would be common for a single shepherd to watch over the whole villages combined herd, which during certain seasons could be quite far from home. As part of the broader colonization, the Spanish introduced their familiar agricultural plants and animals to the hemisphere. Then approximately the same time as settlements form, they released animals into the wild so they could take hold and be available later.

The Spanish herding tradition diffuses to Texas from northern Mexico. The initial release of animals occurs with the Missionaries who introduce Spanish style agriculture. However, systematic ranching really begins with the granting of large land titles. It is these large operations that require skilled workers to tend the animals and produce the meat, fiber, etc. The precursor to the cowboy is the *vaquero* who embodies the same swagger and lifestyle as the Americanized stereotype. In addition to vaqueros, the release and use of horses becomes a major component of ranching as well as life on the settlement frontier. Ample linguistic terms such as lasso, corral, and rodeo survive the arrival of Southern Anglos.

The second component is a Southern ranching tradition. From the beginning of English colonization in Virginia, animals such as cattle and pigs were released into the wild. Because the earliest settlers were concerned with being economically viable, they emphasized crop production that gravitated toward tobacco. At the end of harvest, they would round up animals and separate them by ownership, which required branding. The skills of riding in the tidewater backcountry became highly prized and even wealthy individuals took part in the activities. Yet, the indentured whites and African slaves did much of the ranching tasks. For the Yeoman (whites), the horse riding skills would be useful as they expand westward into Appalachia. It would be slaves that eventually became the main workforce for southern ranching as it develops into a viable economic venture.

The emergence of cattle operations in the Carolinas has both a social and an environmental dimension. Socially, the institution of slavery justified the labor aspect, and the financial freedom that welcomed the input of British investment money was paramount. Environmentally, the Pine Barrens along the coastal plain were suitable for large scale ranching, which did not directly compete with plantations for the same land. As a system, southern ranching diffuses along these marginal lands of the Atlantic/Gulf Coast toward Texas. On the way, it likely was influenced by Spanish and French practices, but it eventually arrived in Texas not significantly different.

It is along the coastal prairie that large numbers of cattle are documented in census records first. The earliest cattle drives followed the Opelousas Trail to New Orleans. In all likelihood, Spanish and southern ranching styles interacted with one another on the coastal prairie as Mexicans and Southerners interacted. The cattle boom in Texas after the Civil War diffuses Texas style ranching throughout the western United States. Two distinct hearth areas of ranching form in Texas, the Cross Timbers (North Texas) and the Nueces Valley (South Texas), from which the diffusion of ranching originates. The Cross Timbers hearth used both black and white cowboys and diffused into central Texas and northward along the Great Plains. Some of the names still associated with this hearth are Chisolm and Goodnight. The Nueces hearth used Vaqueros even after Anglos such as King and Kenedy became the larger landowners. The diffusion of ranching traits from South Texas tends to be further west into New Mexico and beyond. While the large cattle drives were a short lived phenomenon, they helped diffuse ranching as well as images of cowboys and western landscapes. After the railroads constructed railheads in Texas, the adjacent stockyards became important places for trading animals and congregating cultural expertise. Even large cities like Fort Worth took advantage of this new relationship becoming a financial center for cattle futures and eventually an industrial location for meat-packing.

The remaining landscape dimension to ranching is enclosure. With the invention of barbwire, large areas could be fenced in affordably. The law changed to require that ranchers became responsible for keeping their animals in defined areas—their own property or leased land. The theoretical reason for branding to distinguish animals from different herds was eliminated, but the practice remained as a way to detect theft. The impact of fencing has been great. Foremost, the actual fence is an artifact, and the fence line becomes a micro-environment for shrubs and cacti to grow. Visually, fields and property lines are distinguished by the fencing and in many regions intertwined with Prickly Pear and Mesquite.

Part Two of the textbook discussed the historical and cultural geography of Texas. One theme that should have been apparent was the cultural diversity of the state. The past had tremendous variation because of the different migration movements, and we still have similar cultural pluralism today. Another theme was the impact of humans on the landscape of Texas. Human agency is one of the forces that modify the Earth's surface. Today we can look to our cultural landscapes for evidence of the past and indications of the scale of contemporary modifications.

OPEN-RANGE

The reality of wide-open spaces and the mythology of open-range are not the same thing. The landscapes of extensive landuses such as cattle ranching are impressive because of the size of the land units. For example, the areal dimensions to the King Ranch are staggering, and ranch size in West Texas is very large. However, the "range" is not open. Every ranch with livestock is fenced in, and moreover, most of them are segmented internally.

FIGURE 5-23 Advertising Shoot, Culberson County
Photo shoot for Dept. of Commerce near Guadalupe Mountain National Park. Photo courtesy of Texas Department of Transportation.

"Nature Has Spatial Patterns"

FIGURE 6-1 Palo Duro Canyon Landscapes
Palo Duro Canyon State Park, Randall County. Photo by Author.

Part Three

Physical Geographies of Texas

INTRODUCTION TO PHYSICAL GEOGRAPHY

The physical geography of Texas is complex and unique. The complexity partially comes from its size, but also the different physical processes that we find here. The uniqueness of Texas' physical geography derives from the particular combination of processes, which reflects its relative location on the continental landmass. One of the more mentioned processes is precipitation because it seems as if we constantly have too much or too little rainfall. In fact, Texas falls along the longitudinal border between the humid and arid climates in North America. Many early writers of Texas history such as Prescott Webb thought this was a significant environmental feature that contributed to the development of ranching in Texas. Along the other axis, the latitudinal expanse plays a role in the range of temperatures. Equally significant is that the severity of winter conditions also follow a North–South pattern.

In addition to routine weather, climatic processes explain many of our deadly hazards in Texas. On a wider scale, geological processes created the major physiographic regions that underlie all human activity in the state. Geological processes also are responsible for the historical development of hydrocarbons. Texas has a rich diversity of flora and fauna because of its relative location. The way Texans have interacted with the environment and utilized natural resources influences the perception of the physical world. One has only to look at oil and water to see clues about those perceptions.

Without a doubt, the physical geography of Texas is just as diverse as the cultural geography. A good way to appreciate the physical variety is to be exposed to some of the different physical processes and then to learn the generalized regions based on physical criteria. With this mind, Part Three has three chapters. Chapter Six surveys the major processes and details a few of them with exact data and maps. Chapter Seven describes the physiographic regions of Texas. Finally, Chapter Eight explores the human-environment relationship in Texas.

PHYSIOGRAPHY

Physiography is literally defined as the description of nature. A more inclusive dictionary would add that the description includes a class of objects and physical realms. The Earth's surface is a class or realm where humans live and interact with nature. Yet humans are also the narrators of this description. Therefore, physiography is the human-interested description of the Earth's surface. From this perspective, the Earth's surface has some basic features: material, shape, water, life, and human presence. The way one describes these features varies from dispassionate science to romanticized fiction.

Material: The literal material of the Earth that we stand on is the first feature. What is the composition of this material? Is it stable or unstable? What qualities especially fertility does the soil have? There are times we can sense the colors and textures at the surface, meanwhile other times we need to dig down and examine the strata or layers beneath. Broad, regional patterns exist because the scale of geological origins is immense, but at the same time, local specifics exist because of the current geomorphic and biologic circumstances.

Shape: The topography of the Earth's surface or the lay of the land is always apparent. Topography, on and off the map, includes elevation, slope, and aspect. Highlands or uplands are at the higher elevations; lower elevations are the lowlands and bottomlands. A nearly level or flat surface is typically called a plain. The opposite scenario of large slopes and steep terrain is typically called mountains. Hills are in between and somewhat relative to either perspective. Aspect is the direction of the slope and it has special emphasis with sunlight and water flow.

Water: The presence and proximity of fresh water is a crucial factor in human settlement. Biologically, we require water to survive; furthermore, we use water for transport and energy. Even in today's modern landscape, we remain connected to the supply and delivery of water. Water is ever present in human descriptions of nature. Because we cannot unravel people from water, rivers, lakes, and adjacent seas are part of our natural geographies.

Life: The biological communities or ecosystems are another key aspect of a description of nature. For many it is what they sense first: the color of the field, the scent of the forest, or the sound of an animal. The plants and animals that are present is the biological realm, which can become a guide for future ecosystems.

Presence: Anthropogenic change and legacy is also part of a description of nature. Human interest is always part of the mindset for authors. One senses and observes what other people have already done to the environment. In addition, those elements of nature that we think are most important to humans are more often described such as food and water sources.

LEARNING OBJECTIVES: PHYSICAL GEOGRAPHY

Part Three of this textbook is associated with physical or natural elements of Texas enviroments.

- Describe the physical geography of Texas.
  You should be able to describe statewide patterns of major phenomenon and know the general trends beyond the state's borders.

- Locate and elaborate on the major physical regions of Texas.
  You should be able to identify these regions on different maps and know which ones extend beyond the borders.

- Describe the physical geography of your hometown.
  You should be able to identify the major physical patterns and regions for your hometown.

- *Maps:* Familiarity with physical maps of Texas.
  You should be able to read data from maps for your hometown and campus as well as any extreme data locations.
  You should be able to recognize physiographic and climatic maps and their patterns.

- *Definitions:*
  Physiography/physiographic
  Geomorphology
  Climatology
  Hydrology
  Biogeography
  Topography
  Natural resource
  Natural hazard
  Fossil fuels
  Pollution
  Aquifer
  Reservoir
  Extra-Tropical Strom

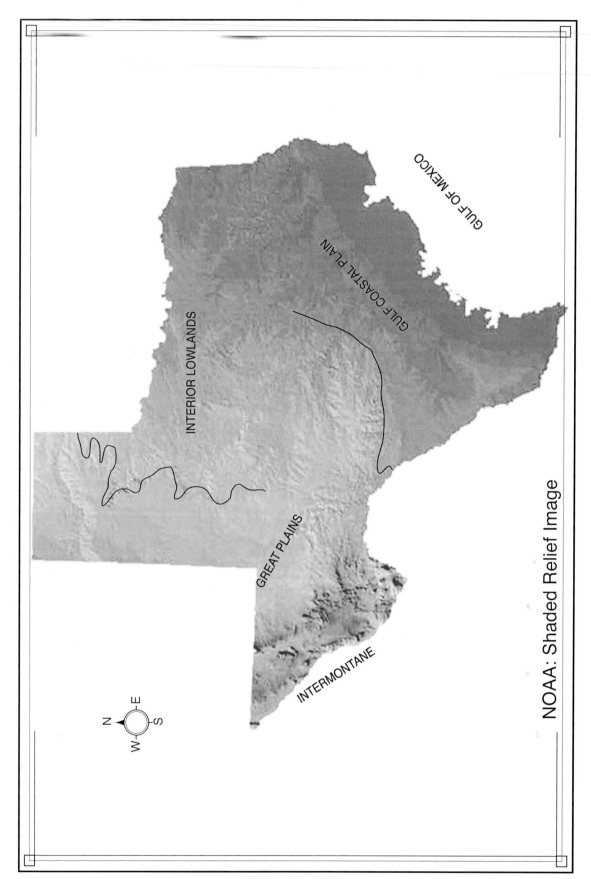

FIGURE 6-2 NOAA Relief Image of Texas
Source: National Oceanic and Atmospheric Agency.

6
Physical Processes

Several different physical processes influence the geographic patterns that we observe in Texas today. In reality, the whole planet is a complex physical system with multiple interactions between the components. When a change of physical state occurs in one place, it could theoretically set off a sequence of events that impacts all other places on the planet. This complete Earth system includes movements of energy and mass from place to place which can be empirically observed. While we know that everything is interconnected, it is useful for scientists to distinguish the parts for separate analyses. In addition to measuring and recording process related data, the existing arrangement of energy and material is recognizable as landforms.

Logically, the systematic study of the Earth began with visual observation, so comprehensive guides were written that documented as much as possible. Progressively, the studies became more scientific and universal in terms of applicability. Today, the pursuit of rule making at a single location goes hand in hand with the big picture of global change. For example, the understandings of sand transport along beaches and fire history of forests are theoretically linked with recent concerns for global climatic warming. In this chapter, the discussion turns to the various processes that shape the physical world around us.

EARTH SYSTEM

- Lithosphere is the innermost part of the Earth. While we think of it as the solid earth (terra firma) at the crust, the material in the core and mantel can be anything but solid at the extreme pressure and temperature found there. Moreover, plates of the crust move in relationship to one another.
- Atmosphere is the gaseous realm that surrounds the other spheres, and from a planetary perspective is the true surface. Nitrogen and oxygen are the most frequent gases, and with the other gases are also found in the immediate land and water environments. The atmosphere is very mobile and is important to understanding transport of energy around the planet.
- Hydrosphere is the realm of water, which overlaps with the other spheres. Much of the planet's water is in the oceans, however, water in all its states, solid ice, liquid water, and gaseous vapor, is found almost everywhere. Fresh drinkable water is much more circumspect in terms of percentage and location.
- Biosphere is the realm of life. Living organisms in all their variety are found in a narrow band close to the Earth's surface. Terrestrial humans, aquatic plants, and flying animals demonstrate the overlap with the other spheres, yet much of the life on our planet is microscopic and undocumented.

GEOLOGIC PROCESSES

The geological processes that shape the Earth's surface are ongoing and constant. These processes range from the spectacular volcanic eruptions and sudden earthquakes to mundane transport of sediments and never ending waves at the beach. Yet to really grasp the physical reality one must appreciate the temporal dimension that implies what we see happening around us today like erosion and deposition also occurred millions of years ago. And other activities such as mountain building and subsurface compaction are not perceptible but they occur as well. While the basic processes were similar, the actual land/water surface of continents and oceans was different during previous time periods because of plate tectonics and sea level changes.

Geological interest in Texas has been quite intense because of oil exploration, yet the state has a wide variety of phenomena to sustain academic and layperson's interest. All the basic types of rock are found in Texas: igneous, metamorphic, and sedimentary. Moreover, the time frame of the surface ranges from the extremely old Precambrian to literally the present as new surfaces are being created. Of course there are spatial patterns to the geology. Figure 6-2 is a map with some of the essential geological features of Texas.

S-curve line

The first thing to note on Figure 6-2 is the bold "S-curve" line that divides the current state of Texas into two distinct sections. The line has multiple associations. Initially, the S-curve line approximates a historical continental margin. Northwest of the line is extremely old base material that some scholars call the *Texas Craton*. Cratons are the stable core of continents that move as a single unit during plate tectonic movements. They have been relatively immune from the dramatic events along the continental margins. In many instances, cratons are covered by sediments that somewhat hide their age and existence. Southwest of the S-curve line was a prehistoric ocean, and immediately offshore was a deep trench. The sediments that deposited in this trench were transformed into above surface mountains approximately 300 million years ago.

Secondly, the S-curve line represents the Ouachita Mountain Range. As part of the collision between the nascent North American and European/African plates, mountain ranges developed on both plate edges. The Appalachian and Ouachita Mountains are the expression of this tectonic activity in the Americas. On the northwest side of the mountains, water and sediment moved inland that deposited in the area we now call the Permian Basin. On the southeast side, the long process of deposition that creates the coastal plain begins. While the Ouachita Range has been denuded and covered over with sediments, the metamorphic rock remnants are mostly subsurface; at Marathon and Solitario in the Big Bend area the range does become exposed at the surface.

The third association of the S-curve line is the Balcones Fault Zone (BFZ). Knowing that the line was already a significant geological boundary, it is not surprising that there is still some faulting. When the Edwards Plateau was uplifted about 20 million years ago, a series of faults or breaks developed along the line. The name "Balcones" comes from this stair-step appearance of the landscape that properly goes by the term escarpment. The nature of the fault exposes the limestone material of the Edwards Plateau as well as the lateral edge of the Edwards aquifer which allows for numerous natural springs.

Surface Expressions

Much of the state is covered by sediments of various ages. The general pattern is a northwest to southeast one corresponding to the topographic relief between uplifted mountains and the Atlantic Ocean. Typically, new layers of sediment wash over previously deposited layers, so the strata are youngest at the surface to oldest at depth. The build-up of material along the Gulf of Mexico shoreline reveals younger surfaces

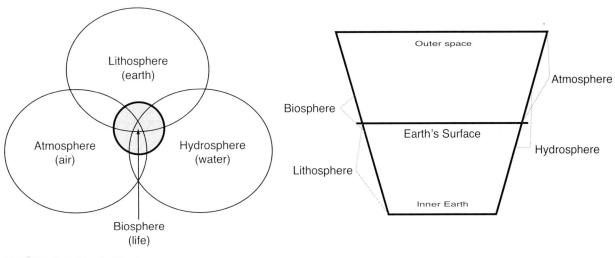

FIGURE 6-3 Earth System

closest to the Gulf. However, inland areas that are currently exposed to erosion might have exposed older material at the surface.

Three broad areas of the more recent Cenozoic materials can be discerned for Texas. First, nearly the whole coastal plain is Cenozoic, and along the coastline there are extremely recent Quaternary materials. Secondly, the High Plains of the Panhandle are covered by Cenozoic sediments that washed eastward from the Rocky Mountains. The third area is the valleys and basins of Far West Texas that receive sediment with the runoff from near by mountains. Cenozoic aged volcanic materials are also found west of the Pecos.

Mesozoic aged materials can be found in a U-shaped pattern with an eastern and southern edge along the S-curve line. The western extent of Mesozoic material stretches up to the High Plains. The interior of the U has been eroded to reveal older Paleozoic material. Further Paleozoic materials at the surface can be found along the Canadian River and elevated areas in Far West Texas. Materials that are extremely old and generally lacking fossil records are known as Precambrian. There are three locations in Texas with Precambrian rock. The largest expanse is in Central Texas around the city of Llano. *Enchanted Rock* is an exposed part of this material that resists erosion and stands above the more level surface around it. The other two Precambrian spots are the east side of the Franklin Mountains in El Paso and the Carrizo Mountains near Van Horn.

CLIMATIC PROCESSES

Texas' climate ranges due to its large area. In many respects, Texas has many different climates (in the plural) depending on where you are. For many people, "climate" is synonymous with weather, and the association is essentially correct. Experiential things such as rainfall, heat index, hard freeze, and drought are communicated by a "weather" person. Oddly enough, the term climate is also used with ones

TABLE 6-1 Geologic Time

Era	Period	Epoch	M.Y.	TX Surface
CENOZOIC	Quaternary	Recent	0.1	
		Pleistocene	2	
	Tertiary	Pliocene	5	
		Miocene	24	BFZ (Balcones Fault Zone)
		Oligocene	37	
		Eocene	58	Volcanism Trans-Pecos Coastal Plain
		Paleocene	66	
MESOZOIC	Cretaceous		144	(Rocky Mtns).
	Jurassic		208	
	Triassic		245	
PALEOZOIC	Permian		286	
	Pennsylvanian		320	Ouachita Mtns.
	Mississippian		360	(Appalachia + Ozarks)
	Devonian		408	
	Silurian		438	
	Ordovician		505	
	Cambrian		570	
PRE-CAMBRIAN	Proterozoic		2500	Central Texas Uplift (est. 1350) Franklin & Carrizo Mountains
	Archean		4500	

Sources: Roadside Geology of Texas (Spearing); Texas Almanac.

perceptions of social environments, so perhaps climate ultimately means how we feel about the conditions at a specific place.

Two distinct academic fields developed around the milieu of atmospheric information. Meteorology studies the science of the atmosphere, and climatology studies the long-term statistical data. While the weatherwoman touts her meteorological society credentials, she mostly reads climatic data measured by static instruments and compared with historical records. Regardless of the academic field, there are some regional patterns of climatic data that are fundamental to know. In addition, there are some universal factors that affect everyone's weather and contribute to these patterns.

Global Factors

Four universal factors stand out in the experience of climate. First, the *latitude* of a place is extremely indicative. The broad pattern of energy from the Earth–Sun relationship correlates into latitudinal temperature patterns. Furthermore, latitude is directly related to seasonality of energy in terms of both sun angle and length of day. The equatorial region has an energy surplus and generally speaking warmer temperatures. The polar regions have an energy deficit and correspondingly cooler temperatures. As a natural system that inherently tries to balance itself, there is a movement of energy from the tropics to the poles via

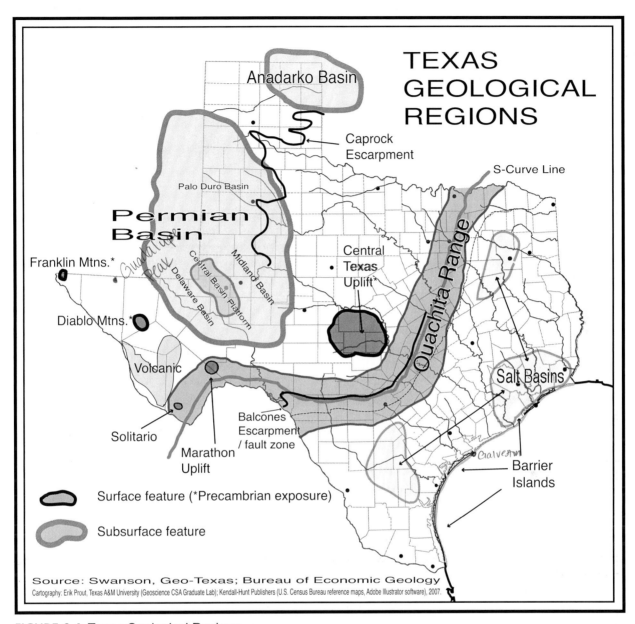

FIGURE 6-4 Texas Geological Regions

both the atmosphere and oceans. For Texas's relative location in the northern hemisphere, we experience a north-south pattern with temperature data.

The second global factor is *elevation.* By itself, the atmosphere changes with altitude, and air that rises or descends changes its characteristics. On the surface, we experience a change of temperature with a change of elevation. Known as the environmental lapse rate, there is a corresponding temperature change of approximately 3.5°F for every thousand feet of elevation. Therefore, we would anticipate a 30-degree difference between Galveston and Guadalupe Peak assuming no other factors were in play. Generally speaking, higher elevations are found in the western side of the state. The highest elevated city is Fort Davis at 5,050 feet above sea level, which is just slightly less than Denver, Colorado for comparison. Elevation in the form of topography influences precipitation patterns. When moist air ascends over

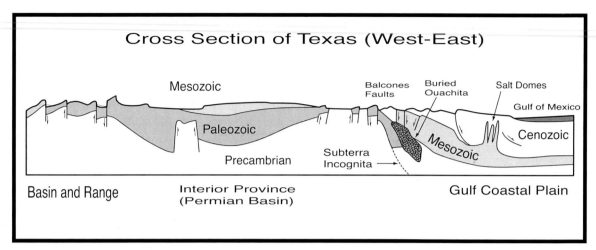

FIGURE 6-5 Geological Profile of Texas
Source: E. Swanson, 1995. *Geo-Texas* (p. 64). The profile of Texas reveals both structural and temporal variety. A general pattern of older material inland of the S-curve line and younger material coastward is repeatedly interrupted.

mountains, there is an increased chance of precipitation of the windward side. Likewise, there is a decreased chance of precipitation on the leeward side as air descends. At a macro-scale, the Rocky Mountains and Sierra Madres create a *rainshadow* on their eastern (Texas) sides. At a micro-scale, the Davis and Guadalupe Mountains have enough topography to register both cooler temperatures and higher precipitation amounts.

Continentality is the third global factor. Continentality refers to a place's relative location on the large landmasses of Earth. The basic principal is that land and water have different thermal properties especially how they heat and cool in response to solar energy. Basically, land heats and cools faster whereas water is slower because it circulates, dissipates, and stores energy better. The larger the landmass, the more likely the seasonal variations in temperature will exist especially near the center. Both Eurasia and North America have monsoonal patterns, and being located where these seasonal winds (and rains) influence weather is paramount. Because the land/water pattern is unique *(Geoid)*, it is generalized into an inland-maritime continuum. Inland areas have larger temperature ranges; coastal areas have more moderate temperature ranges. For Texas, this inland-maritime dimension translates into proximity to the Gulf of Mexico. In addition, the Gulf contributes most of the moisture that translates into humidity and potential for precipitation. Proximity to the Gulf mostly explains the east-west pattern to precipitation.

The final global factor is air mass boundaries and storm tracks. Air masses are parts of the atmosphere that acquire characteristics of the Earth's surface beneath it. Air masses that form over water are called maritime (m) and are moist; those over land are dry and called continental (c). Air masses vary by temperature as well; tropical (T) and polar (P) are the two common modifiers. By combining the two dichotomies we end up with four basic air mass types: mT, mP, cT, and cP. Eventually, air masses move away from their origins and come into contact with other air masses. All four of the idealized types influence Texas' weather in at least one way or another. The maritime tropical air mass (mT) comes to us from the Gulf of Mexico as it brings most Texans their summertime humidity. Those hot dry and perhaps drought months are often dominated by the continental tropical air mass (cT) that forms over Mexico. Those cold frigid blasts of air that are called "blue-northers" are continental polar air masses (cP) that descend southward out of Canada. Maritime polar air masses (mP) routinely move across the United States as part of a westerly flow of winds in the mid-latitudes. The "westerlies" migrate seasonally and have their maximum effect in Texas during winter months. One last aspect of these air masses is the contact zone between them called fronts. In North America, the contact between the strongly contrasting mT and cP air masses can produce severe weather; the prevalence of tornadoes is quite significant with names like "tornado alley" becoming recognized.

Temperature

Temperature is a dynamic variable that occurs everywhere spatially and is always present temporally. Yet it can be measured and presented in numerous ways. Unfortunately, the historical record is limited to specific locations, what was actually documented, and the reliability of the instruments and operators. Contemporary remote sensing from satellites can estimate temperature in a more uniform way, but it cannot reconstruct the past. The most common way to record temperature is to document the high and low for each day. Then those two numbers are averaged together to create a daily average. Furthermore, daily averages are calculated into monthly and yearly averages. By the time an annual average temperature is calculated the statistical anomalies are usually worked out, and it's entirely reasonable to say such and such place is warmer than another place.

Temperature patterns in Texas strongly correlate with latitude. The patterns also reflect elevation in a secondary way. On a map it appears as a north/northwest–south/southeast pattern with some anomalies in Far West Texas. The range of annual average temperatures in Texas is between 55°F and 75°F. The highest annual average temperature in Texas is found at the Brownsville in the Lower Valley, which is the southernmost station in the state. The lowest annual average temperature is located in Dalhart County in the northwest corner of the Panhandle. Topography in the westernmost part of the state, especially the Davis Mountain weather station, explains the slight departure from the latitudinal pattern.

Other climatic variables that are related to temperature include freezing dates and growing season length. For every location that records temperature, an average date for the first freeze and the last freeze can be determined. Using those two dates, the length of winter and potential freeze can be calculated. Growing season is the reverse; it is the average length between the last freeze and the first freeze. Farmers have to select and plant their crops with this in mind. While not strictly climatic, energy consumption is closely related to temperature. In Texas, we get this at both ends of the spectrum. Higher latitudes will need more energy in the winter half of the year, and increased consumption is associated with cooler temperatures. Lower latitudes with higher summertime temperatures will need additional energy for air conditioning.

Precipitation

Precipitation is also a dynamic variable that can be measured from radar and satellites. The traditional way is to collect precipitation in a preset container and record the amount per day and/or event. Precipitation includes rain, snow, sleet and hail; if the precipitation falls frozen it can be measured after is thaws. Snow fall and snow cover are also recorded, and for cold weather areas snowfall is translated into a rainfall equivalent. Each day regardless of the number or types of events has a single number. Theoretically, the yearly total amount could be averaged by 365 days, but the common way is to simply record the daily, monthly, and yearly totals. The monthly totals can be averaged to graphically show which months are the wettest and driest as well as any corresponding seasonal pattern. The calendar year totals are averaged to give each weather station an average annual precipitation amount.

Precipitation patterns in Texas have a strong East–West pattern. The longitudinal pattern continues northward dissecting the United States and reflecting a broad humid east and arid west. The lowest annual precipitation average is in El Paso at a measly 8 inches per year. El Paso is distant from both Pacific and Gulf moisture. Port Arthur on the other hand is exposed to Gulf moisture whenever the winds are out of the south. The highest annual precipitation average is in the Southeast with a whooping 58 inches per year. Halfway between these two data extremes, the 33 inch isohyet is located just east of Austin. Much of the state experiences less than 30 inches of precipitation, which some scholars consider to be an indicator

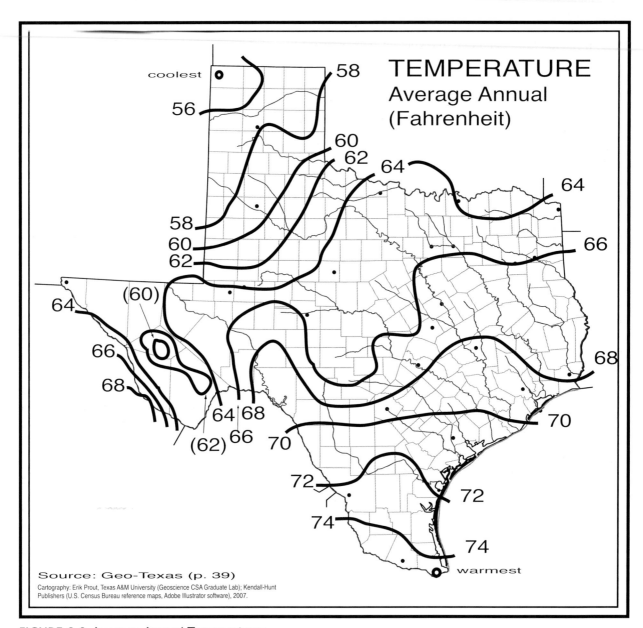

FIGURE 6-6 Average Annual Temperatures

of aridity. A more exact calculation would be to compare precipitation (input) with the potential outputs of evaporation and transpiration. An even more exacting measurement is soil moisture indexes that measure the amount of water in the soil. Generally speaking we still see an east-west trending pattern for these other calculations.

In addition to an average annual precipitation amount, there are two significant dimensions that have a similar longitudinal pattern. The first is seasonality of precipitation. Seasonality refers to what part of the year does the precipitation occur. For easternmost Texas, precipitation occurs year round with roughly half falling in summer and half in winter. El Paso on the other hand receives nearly all its precipitation in the summer half of the year. Variability is the second dimension, and it refers to the deviation from normal. The western side of the state tends to have higher precipitation variability. Places that already receive small amounts primarily in the hottest half of the year are not likely to get their annual averages. Precipitation variability is much lower in the eastern half of the state. Big Thicket receives ample precipitation that falls year round and rarely deviates far from its average total.

Classification

Climate classification is the generalization of different weather experiences into a few categories. It is an old enterprise as exemplified by the Greeks who classified according to latitude. Temperate was their own zone, and they looked down on those people and places both north and south from themselves. Early explorers and settlers in America observed the moisture transition that became ingrained in the popular imagination as the humid east and arid west. As reliable data became available for the whole planet, it was

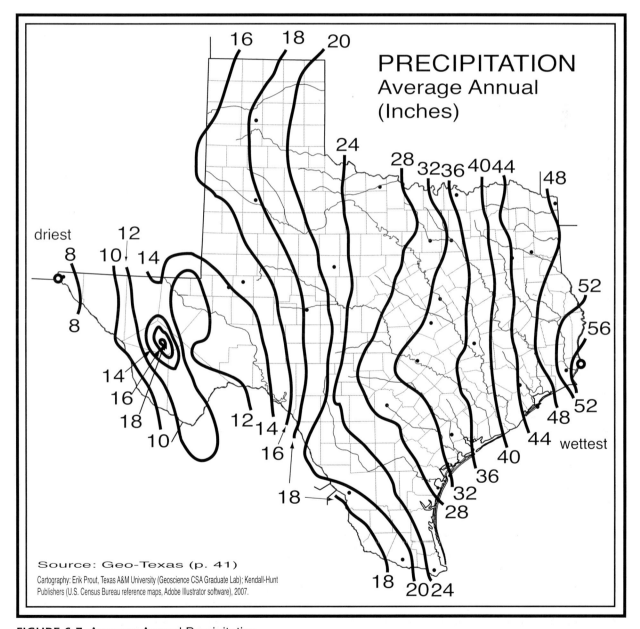

FIGURE 6-7 Average Annual Precipitation

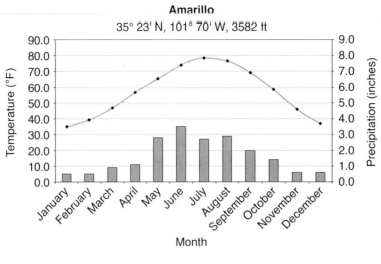
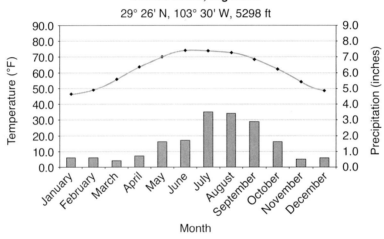

FIGURE 6-8 Select Climographs of Texas Places
Data Source: World Climate.

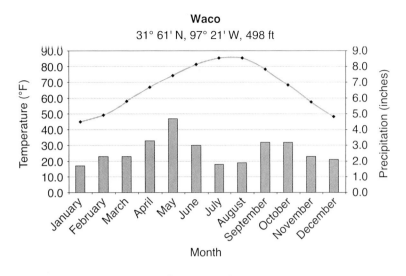
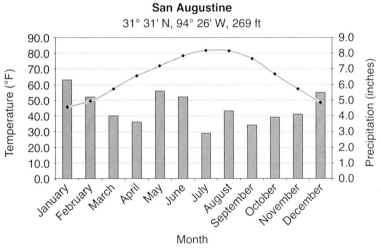
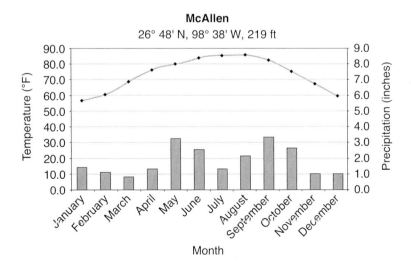

FIGURE 6-8 Continued.

inevitable that a classification scheme would try to generalize everywhere. A German geographer named Köppen devised such a system using temperature and precipitation as the two key variables. Two of the five basic climatic types are found in Texas: Köppen's B and C. The mesothermal (C) climate is one of sufficient precipitation with middle range of temperatures, and it is found on the eastern side of the state. The western side is predominately dry (B) climate with some being differentiation between semi-arid steppe (BS) and true arid desert (BW).

R. J. Russell was really the first scholar to analyze the Köppen system's applicability to Texas. In general, the system worked well, but Russell attempted to fine-tune it. After analyzing Texas' climatic data, Russell thought that variability was an extremely important consideration. The two dimensions of variability act perpendicular to each other. The more nuanced element of the east-west pattern is related to precipitation. Basically, the frequency of drought becomes an overwhelmingly important factor. The severity of winter becomes the finer tuned consideration of the north-south temperature pattern. With these considerations, the Russell modified Köppen climate map for Texas has twelve generalized regions.

Regions 1–7 are all basic C climates with only regions 1 and 2 receiving consistent year round precipitation. Regions 3 and 4 have occasional dry years or moderate droughts, meanwhile regions 5, 6, and 7 are more likely to have extremely dry years and severe drought cycles. Of the basic B climates, all except for region 12 are classified as steppe, but they all have extreme drought cycles. All of the B climate regions (7–12) in addition to regions 5 and 6 tend to have dry winters (w) with most precipitation arriving in the summer half of the year. Regions 7 and 11 are labeled with a "CA" to distinguish their occasional warm (tropical) winters. Regions 1, 3, 5, and 9 are labeled with a "CD" to distinguish their occasional cool (microthermal) winters. Region 8 has an inverted "DC" to signify that they have a cool winter over half of all years.

HYDROLOGIC PROCESSES

Water is an essential component to life on our planet. Humans are very sensitive to drinkable water supplies, so historically human settlements had an obvious relationship to freshwater lakes and rivers. With technology and wealth, humans can transport water long distances, pump water from deep under ground, and even build cities in the desert. However, humans cannot really create water except in very controlled laboratory conditions. One of the tenets of *hydrological cycle* is that water is neither created nor destroyed. The hydrological cycle is a systems approach to water that tracks its movement, change of state, and storage. Schematically, water evaporates from the oceans to become vapor in the atmosphere. Then, some of this water vapor moves over land and falls to the surface as precipitation. Finally, water flows downstream back to the oceans completing the cycle.

The overwhelming majority of water on Earth is located in the oceans, which is too saline for human consumption. The largest amounts of freshwater are locked up as ice sheets in Antarctica and Greenland. Smaller amounts of freshwater are stored in Alpine

TABLE 6-2 Global Water Percentages

Location/State	Percentage
All Water:	
Saline water (Oceans)	97%
Fresh water	3%
—available for humans	0.3%
Fresh Water:	
Ice caps and glaciers	69%
Ground water (aquifers)	30%
Surface water (rivers, lakes, etc.)	0.3%
Atmosphere	0.04%

Source: U.S. Geological Survey.

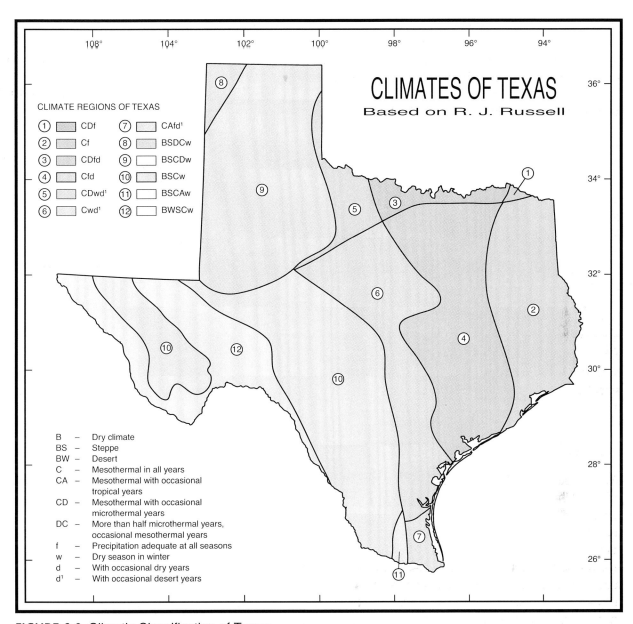

FIGURE 6-9 Climatic Classification of Texas
R. J. Russell's initial classification of Texas climate. Previously published in the Annals of the Association of American Geography.

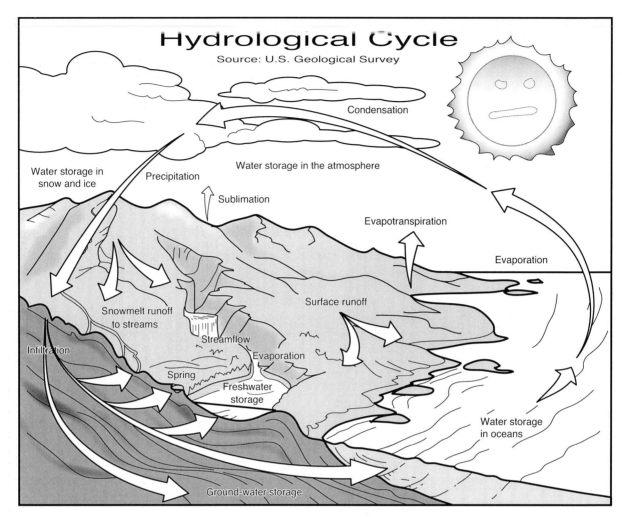

FIGURE 6-10 Hydrological Cycle
Source: USGS.

glaciers at higher elevations as well as the seasonal melting of snow pack. Rivers that are fed by snow and ice melt are extremely important for the societies that formed downstream, for example, the Nile and Rhine Rivers. In addition to rivers, fresh water can enter the ground where it can move and be stored for long periods of time in an aquifer. An aquifer is a water-bearing strata of mostly subsurface rock. Typically, there is a part of the aquifer that is exposed at the Earth's surface where recharging occurs. Broadly speaking we can conceive of two types of water: surface and ground.

Surface Water

Surface water flows in predictable ways based on the quantity of water and shape of the terrain. On many maps, surface water or fluvial systems are drawn as single lines to locate and label the main stream flow of a river. However, surface flow is possible everywhere, so river systems actually consist of numerous branches or tributaries that drain an area. These areas are properly called drainage basins; the boundaries between basins are divides. Rivers follow fluvial properties associated with moving water including the ability to erode, transport, and deposit material. Each river can have statistical data associated with it such as length (main channel), basin area, discharge (amount of water), and load (amount of sediment).

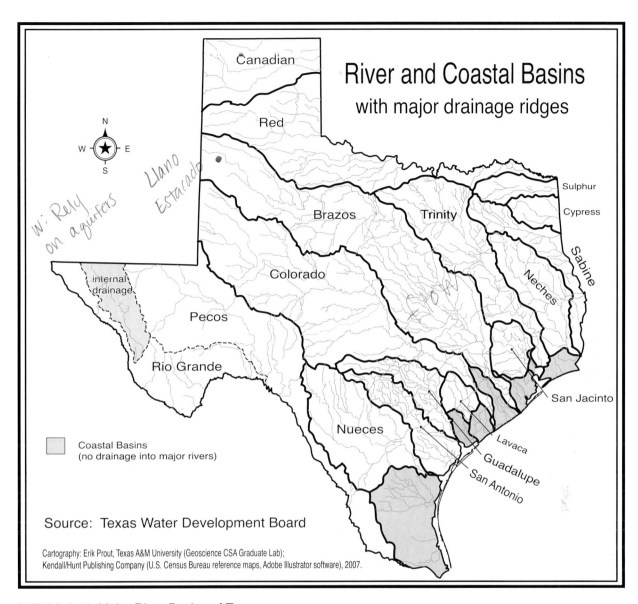

FIGURE 6-11 Major River Basins of Texas

There are 12 to 18 major river systems depending on how one defines major. Sometimes there are major tributaries that contribute as much discharge as the named stream. The system employed here is to list the cartographically prominent rivers that enter or exit Texas along its borders. Then, major tributaries are discussed as part of those rivers. The general pattern for Texas rivers is that they flow in a southeasterly direction towards the Gulf of Mexico. Most of Texas' rivers originate on the plains and plateaus in the state or on the New Mexico side of the Llano Estacado. The exceptions are the Rio Grande, which originates in the San Juan Mountains of southern Colorado, and the Pecos and Canadian Rivers on the east side of the southern Rockies in New Mexico. The Canadian and Red Rivers are part of the greater Mississippi River system. On the other hand, all the other rivers terminate directly along Texas' Gulf coastline.

Ground Water

Water at the Earth's surface can evaporate back to the atmosphere, flow to lower elevations, or percolate into the soil. Once water is in the topmost layers, it can move laterally or infiltrate further down. Conceptually,

TABLE 6-3 Texas Rivers

River	Length (miles)	Length (Texas)	Basin Area (sq. miles)	Area (Texas)	Discharge (acre-feet/year)
Canadian	906	213	47,705	12,865	196,000
Red (Wichita)	1360	695	93,450	24,297	3,464,000
Sulpher	222	222	3767	3580	932,700
Cypress	90	75	3552	2929	493,700
Sabine	360	360	9756	7570	5,864,000
Neches (Angelina)	416	416	9937	9937	4,323,000
Trinity	550	550	17,913	17,913	5,727,000
San Jacinto	85	85	3936	3936	1,365,000
Brazos (Navasota)	840	840	45,573	42,865	6,074,000
Colorado (Llano Concho)	865	865	42,318	39,428	1,904,000
Lavaca	117	117	2309	2309	227,000
Guadalupe	409	409	5953	5953	1,422,000
San Antonio (Medina)	238	238	4180	4180	562,700
Nueces	315	315	16,700	16,700	539,700
Rio Grande	1896	889	182,215*	49,387*	645,500
Pecos (Rio Grande tributary)	900		44,259		
TOTAL		80,000		263,513	

Sources: Texas Water Development Board; Texas Almanac. Major tributaries are listed in parentheses. The Sulpher and Cypress are technically tributaries of the Red River; the Pecos is a major tributary of the Rio Grande.

*Includs Pecos River area

ground water has two dimensions. The first is a local water table that rural users need to know about so they can dig wells for access. The second dimension is the more regional scale of aquifers. An aquiclude is a non-porous material, and an aquifer is a layer of porous material that allows water to move. Often sedimentary layers alternate between aquifers and aquicludes, so water-bearing layers are confined and limited to where they can be recharged. Approximately, 80 percent of the surface area of Texas has a viable aquifer underneath it. Moreover, the relationship is even stronger for the population distribution.

The cartographic patterns of aquifers correlate with geological areas. The Ogallala aquifer underlies the Great Plains. A series of aquifers parallels the Gulf of Mexico coastline. Aquifers in Texas can be exceptionally deep because they are associated with sedimentation processes along the changing coastline. Texas acquires a majority of its water from aquifers, yet there is some differentiation between users. Approximately, 40 percent of municipal water use comes from aquifers; whereas 80 percent of agricultural use derives from aquifers. As a rule of thumb, communities in the western half of the state rely on aquifers more than the eastern half.

BIOGEOGRAPHIC PROCESSES

The biogeography of Texas is incredibly diverse. The life forms found around the state vary because of the different environments. One would expect numerous environmental niches for plants and animals because of its size. Just as significant, Texas' relative location puts it at a crossroads of large natural realms that cover the continent. The first of the big realms is the Chihuahuan Desert of Northern Mexico.

FIGURE 6-12 Retaining structure along Llano River
Llano River in Llano, Llano County. Photo by author.

The expanse of plains and prairies across the middle of the United States is the second. Thirdly, the forests of the American southeast extend into East Texas. While humans are an integral part of life on this planet, we'll set them aside for now and discuss the other elements.

Vegetation

Vegetation is an obvious component of landscape because it is one of those elements that we actually observe as we go about our lives. For example, trees, grasses, and crops are almost everywhere. Sometimes, vegetation is so ubiquitous that we may overlook it or assume it is natural: lawns and landscaping are prominent examples. Other times, vegetation is so overwhelming that you cannot overlook the fact of being in the forest or on the prairie. Vegetation can also be observed and studied from above with remote sensing. In fact, the preponderance of land cover is vegetation in one form or another, and the technical challenge is to distinguish vegetation types. Unfortunately, vegetation maps can be extremely complex, and the base map for Texas has over 90 generalized categories.

A further issue is whether vegetation is natural. Many scholars want to understand the natural vegetation patterns because they are related to other physical phenomenon. Unfortunately, human modification has been rather extreme with the deliberate and accidental introduction of exotic species, the use of fire, and the organized patterns of agriculture. Some scholars estimate that only about 17 percent of the world's flora is still natural. Therefore, in many regards natural vegetation is a theoretical concept that asks which vegetation (and ecosystems) would exist without human manipulation. Yet, many of the introduced crops, for example, fill a similar environmental niche as the natural plants. The final issue is to explain the patterns. Broadly speaking, vegetation patterns are most closely related to climatic patterns. Plants are limited to specific temperature and precipitation ranges. In Texas, we find both a north-south and an east-west dimension to natural vegetation. A second factor determining vegetation patterns is geological. Soil chemistry is particularly important in Texas where too calcareous soils prohibits woodland and causes grasslands. A final consideration is the role of the immediate ecosystem. Certain types of vegetation require another species or a coherent ecosystem to thrive.

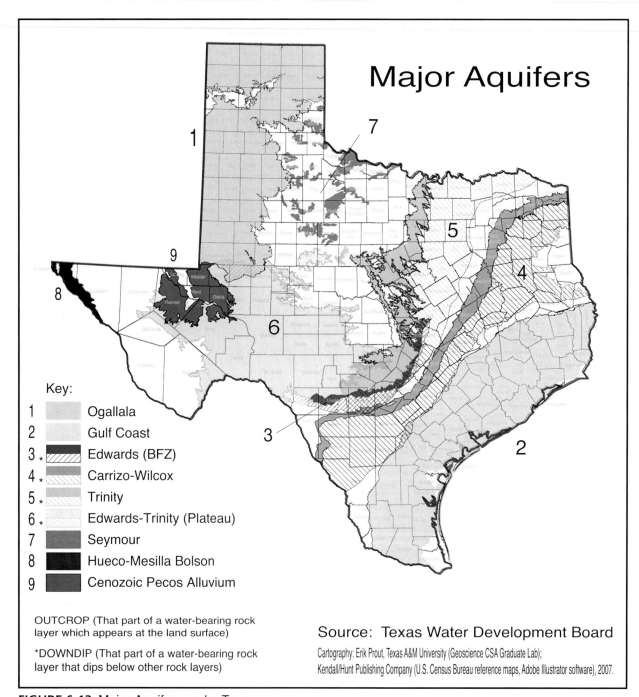

FIGURE 6-13 Major Aquifers under Texas

Piney Woods

Of the dozen or so generalized vegetation regions, the Piney Woods of East Texas stands out as both a physical and cultural region. Foremost, the Piney Woods are dominated by dense stands of various pine trees—notably the loblolly pine. There are other softwood species as well as hardwood species, such as Hickory and Cypress, especially along waterways. The coniferous forest (cone bearing) of Texas is a continuation of the southeastern forest complex that extends to Georgia. This southern forest has been used as

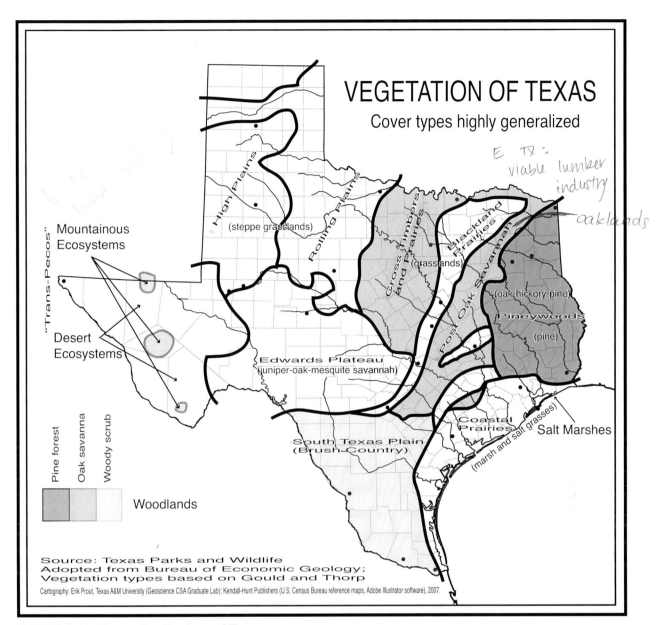

FIGURE 6-14 Vegetation Regions of Texas

a resource for Centuries. Initially, it was a source of "Naval Stores" and later large scale forestry. Much of the Piney Woods has been harvested—typically clear-cut, and what we see today is replanted species. East Texas maintains a viable lumber industry based on careful forestry management and contemporary harvesting and milling techniques.

Brush Country

The South Texas Plains are often described as having dry, subtropical vegetation. Much of the region should actually be classified as chaparral woodland because most of the species tend to have woody stems, yet they are stunted due to less than ideal growing conditions. Also known as Brush Country or *Monte* in

Spanish, the region does have its shrubs, prickly pear cacti, and mesquite. In more favorable areas, live oaks and post oaks take hold, and numerous grasses have been reseeded to augment cattle ranching.

Coastal Marsh

Along the immediate Gulf coastline, a marsh vegetation complex exists. Generally speaking, the coastal prairie and marshes are very flat and wet with undeveloped soils. The tidewater complex consists of marsh grasses and salt grasses that tolerate seawater intrusions. The wetlands continue inland where a coastal prairie has developed with long and medium grasses that are ideal grazing conditions.

Oaklands

West of the Piney Woods, hardwood oaks become the dominant tree species. A post oak savanna, which includes grass species spread about the woodlands, stretches from the Red River down to the lower Guadalupe River. The post oak is the most common tree, but blackjack oaks, live oaks, and elms are also numerous. Further west, the Cross Timbers has a similar mix of tree species with more invasions of mesquite and junipers.

Prairies and Steppe Grasslands

Grasslands covering the open plain are one of the stereotyped images of the state. Texas has numerous natural grasslands, yet some have their share of woody vegetation especially along waterways. The northern prairies (Black and Grand) had medium and large grasses before agriculture. The rolling plains and high plains tended to have a short grass steppe environment. Much of the grasslands have been used for ranching, so high quality grazing grasses have been introduced.

Far West

The Trans-Pecos consists of a very complex pattern because of the topographic variety and very different environments found there. Two generalized ecosystems, each with their own vegetation types, are in the far western side of the state. Desert vegetation that is dominated by cacti, yuccas and other arid species is located at the lower elevations. At higher elevations, western mountain ecosystems exist with piñon pine, juniper, and ponderosa pine.

Anomalies

A few small regions are worth mentioning with vegetation. The first is *Lost Pines*, an area of pine forest near Bastrop. Lost Pines is a remnant pine forest that survived as coniferous forest retreated eastward to present day East Texas. *Lost Maples* is a remnant hardwood forest located north of Uvalde. With a similar explanation, the ecosystem survived in a unique niche despite the changing patterns around it. In the arid deserts of the West, natural springs create islands of oasis vegetation. Balmorhea Springs and Hueco Tanks are good examples of what water can do in the desert.

Wildlife

The zoogeography of Texas is very complex and extremely dynamic because animals move. Some species migrate and are only in the state for a limited amount of time. A good example is birds; Texas is located along one of the major fly-over routes for north-south migrations. Over 540 different species of birds have been identified in Texas; a good number of them along the Gulf Coast as part of the fly-over routes. The variable environmental niches allows for a wide range of species not only of birds but animals. Texas has

FIGURE 6-15 Wildlife of Texas
"Meet Texas' official State Small Mammal, the armadillo." (Previously published in *Texas Highways*.) Seagulls, Padre Island National Seashore. Photos courtesy of Texas Department of Transportation.

FIGURE 6-16 Riparian vegtation along streambed
Lost Maples State Natural Area: Sabinal River, Bandera County. Photo by author.

FIGURE 6-17 Deer along the roadway insided Protected area
Hunting is big business in Texas; Colorado Bend State Park, San Saba County. Photo by author.

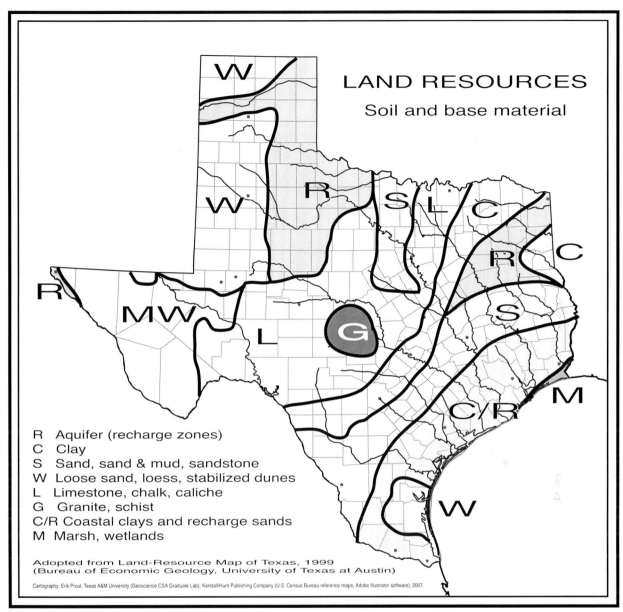
FIGURE 6-18 Land Resources of Texas

over 142 species of animals including 34 mammals. With federal Endangered Species Act provisions, lists of endangered and threatened species are kept by the Texas Parks and Wildlife Department. Texas has its share of species that are endangered and threatened including bats, turtles, woodpeckers, and wildflowers.

The geography of wildlife is difficult to represent on maps because animal territories are not inscribed in the landscape. Most research doesn't try to produce exact counts or precise sighting locations; mapping of species closely resembles habitat or ecosystems that they live in. The larger pattern of animals in Texas follows the big three realms to some degree. From northern Mexico species such as armadillos and jaguars extend northward into the state. The southern forests bring raccoons, squirrels, and deer to name a few into Texas. The Great Plains contribute prairie dogs and historically bison. In a more limited fashion, the coastal marshes of Louisiana sustain alligators and nutria that spread into Southeast Texas. A variety of western species are found at higher elevations of the Trans-Pecos.

Soils

Soils are a significant resource for people because they are the actual Earth's surface that we grow our food in and build our homes on. Soils are also very complex with components of all four of the Earth's spheres: material from the lithosphere, gases from the atmosphere, water from the hydrosphere, and organics from the biosphere. The edaphic processes that create soil include time which is beyond an individual's observation. Early settlers to Texas already had practical knowledge of soils, and they quickly determined where the best agriculture would be located. Some of the variables of soil include depth, color, texture, and fertility. Terms like black(-lands), red(-lands), sandy, and claypan exemplify this appraisal of soil that sometimes becomes part of a place or region's name.

Broadly speaking, the patterns of soil are related to geology of the base material, the climate, and the existing vegetation. Because soil is a fragile resource, soil surveys have been conducted mostly on a county by county basis. The result is that over 1,300 soil types have been documented. Therefore, the generalizations necessary to make statewide "soil" or "edaphic" maps are not typically done. Many government agencies incorporate soils into land use or land resource areas.

In this chapter, physical processes were discussed that literally created Texas and the landforms and landscapes we see around us today. The long timeframe of geology explains the geographical patterns of topography and watersheds. Climatic factors that strongly differentiate the state contribute to the patterns of vegetation. Water and wildlife have their patterns based somewhat on small and large scale factors. As we mentioned at the beginning of the chapter, everything is interconnected to everything else, and every place is somehow related to every other place. Numerous locations and regions were mentioned in this chapter. In the next chapter, a framework or hierarchy of physical regions is discussed.

FIGURE 6-19

FIGURE 7-1 Steppe Grassland along Caprock Escarpment
Palo Duro Canyon State Park, Randall County. Photo by Author.

7
Physiographic Regions

The possible physical regions of Texas are numerous because of the multitude of processes. In addition, Texas' size sets the stage for many locally scaled interpretations of the physical world. One way to organize all the physical regions is hierarchically. In this organization, which works for human regions as well as physical, we conceptualize smaller regions being embedded in larger regions. While the landmass of North America is quite large, physical scientists can generalize this enormous area with all its variation into a dozen or so major provinces.

NORTH AMERICAN PROVINCES

Four major North American physiographic provinces are found in Texas and some scientists would even argue for a fifth. The four provinces are the Intermontane Basins and Plateaus, Great Plains, Interior Lowlands, and the Atlantic and Gulf Coastal Plain. The fifth region sometimes mentioned is the Rocky Mountains. The Rocky Mountains are part of a larger pattern of mountains on the western side of the Americas, and for the United States, coincides with the Atlantic/Pacific Continental Divide. In Mexico, we see this mountainous pattern as the Sierra Madres, yet we see very little surface expression of these mountains in Texas. While geological continuity exists, the surface of the westernmost part of the state is more indicative of the Intermontane even farther West. The Intermontane literally refers to the area between the mountains, which means in this instance the Rocky Mountains to the East and the complex of mountain ranges closer to the Pacific Ocean. In the middle of the Intermontane is the Great Basin that currently has no drainage outlet to the oceans. Two massive plateaus on either side of the Great Basin are the Columbia and Colorado respectively that currently have important river systems that drain to the Pacific. Reflecting the East–West tectonic forces, the Intermontane has a series of fault blocks known as *Horst and Grabens*. At the surface, they appear to be parallel basins and ranges running North–South. In the more drier and warmer south and central parts, the landforms are decidedly arid. Arid landscapes are constantly impacted by Aeolian (wind) processes, but ironically they are shaped dramatically by the rare precipitation events that produce quick movements of material during flash floods. In general, the basin and range topography is often easy to identify because they are not worn down enough to mask their origins. The Intermontane and more broadly the Interior West is the least densely populated region in the conterminous United States.

East of the Rocky Mountains is broad plain that stretches from Alberta Canada to the U.S.–Mexico border, which receives the mountain's eastward runoff of water and sediment. The physiographic term "plain" is used in the United States to label this level region with the specific modifier "great" to emphasize its immensity. Many of the earliest American explorers and settlers called the Great Plains the Great American Desert because it was vegetated with grasses instead of forests. In Canada, the French term "prairie" is used to describe this region emphasizing the dominance of grassy vegetation.

Previously, the Great Plains' prairie grasses had supported large buffalo herds that Native Americans could use for subsistence. After the various Indian Wars, which also decimated the buffalo, American settlers introduced mixed farming to the Great Plains. In retrospect, the settlement was less than ideal with too many people and too intensive agriculture that eventually failed during the Dust Bowl years. Since the 1930s, population has been declining, and agriculture has evolved toward extensive land uses such as wheat and ranching. However, in those areas with irrigation water such as Lubbock, valuable types of agriculture returned.

The Interior Lowland of North America is a vast expanse of the overall continental landmass that coincides with the significant drainage basin patterns. The Mississippi River that includes the Ohio and Missouri and St. Lawrence that drains the Great Lakes occupy this interior section of the continent. The elevations are relatively low when compared to the mountain systems to the East (Appalachia) and West (Rockies). In general, the region is worked over or shaped by these well developed river systems as well as glaciation in the northern half. The Interior Lowland is an agricultural breadbasket for the United States and Canada. It simultaneously supports continuous rural settlement with intensive agriculture and large cities with industrial activities. The rise of manufacturing in the interior was possible because those well developed river systems served as transportation corridors. The Appalachian Mountains serve as a barrier between the Interior Lowland and the Atlantic Coastal Plain. While the Appalachian system continues southwestward and makes a surface appearance with the Ozarks/Ouachita in Arkansas, we usually don't consider it to be a barrier in Texas.

The Atlantic and Gulf Coastal Plain is a continuous plain that extends from approximately New York City down the Atlantic seaboard to Florida and then along the Gulf of Mexico to near Tampico, Mexico. The Coastal Plain is part of the immediately adjacent continental shelf. The level slopes that we see at the surface today extend out into the nearby oceanic waters; therefore, the water is relatively shallow immediately offshore. With changing sea levels over geological time, the exposed or emerged percentage of the Coastal Plain also changes. Sometimes the currently exposed areas were under water, and at other times, the submerged areas were exposed. The Coastal Plain is heavily settled with major cities that served as the major gateways for trade between the United States and the Atlantic world. It also has a contiguous rural population that in some instances date back to original English settlement. The Coastal Plain has experienced numerous changes in terms of labor and crop types; examples include the demise of slave based plantations and the boll weevil decimating cotton production. Agriculture is still widespread on the Coastal Plain; meanwhile forestry is another commonly found primary activity. While the Coastal Plain is the major pathway to Texas, the physical regions are discussed in a West to East pattern starting in the Intermontane and concluding at the Gulf.

INTERMONTANE BASIN AND PLATEAUS

The Intermontane occupies the westernmost part of the state as well as extending out to eastern California and up to Idaho. With this in mind we can use the two political borders, upper Rio Grande and part of the 32° N latitude line, as convenient locators. The eastern border with the Great Plains is slightly vague because we cannot use the "between-the-mountains" definition that the Intermontane term implies. The border is drawn typically on the western edge of the Stockton Plateau and Toyah Basin, which are considered to be part of the Great Plains. For many Texans, "Trans-Pecos" describes the far western part of the state. From the perspective of most Texans who live in the eastern half of the state, "trans" means beyond or the far side of the Pecos River. While this is relatively accurate, it is too much of a generalization when considering the physical geography. In fact, the Intermontane is extremely complex and diverse; it contains both young and old, and topographic relief. The only large place in this physical region is the city of El Paso; yet Big Bend is a well known place, just not known for lots of people.

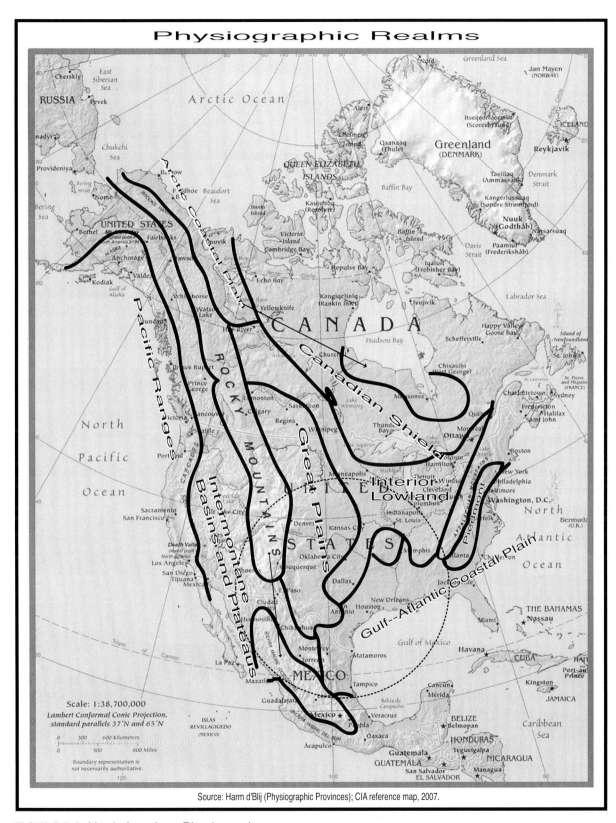

FIGURE 7-2 North American Physiography

Basin and Range Province

Nearly all of the Intermontane in Texas can be subsequently classified as a Basin and Range Province. The dominant physical pattern is a series of parallel north–south basin and ranges. The Delaware Mountains are a good example of basin and range topography. A secondary pattern is volcanic flows such as the Davis Mountains. In addition to basin and range topography, there are small plateaus and widespread recent build up of sediments as well as a salt lake. Since the last major activity 35 million years ago, the landscapes of the province are dominated by erosion. Because of the variation, each of the sub-regions has its own origins explanation.

Guadalupe Range

The Guadalupe Range also runs north–south but its origins are oceanic with much of the surface material consisting of reef material. According to environmental scientists it is part of a broader Southwestern ecoregion classified as the Arizona/New Mexico Mountains. A majority of the range is in New Mexico, but the southern end of the range crosses the 32° N latitude line in Culberson County. The highest recorded elevation in Texas is located at the top of Guadalupe Peak that measures at 8,749 feet above sea level. There are five other named peaks over 8,000 feet high including the oft photographed southern tip, El Capitan. Guadalupe Mountains National Park is located in both states, and recreational facilities can be accessed from the Texas portion.

Davis and Marathon

The Davis Mountains are volcanic in origin dating back 35 millions years, which makes the surface age younger than much of the province. There is enough elevation in the topography of the Davis Mountains to modify the ecological patterns. Basically, precipitation increases enough for a woodland environment to exist. Farther south near Alpine, Paisano is a collapsed volcano. The Marathon Uplift in the Big Bend area is actually part of the Appalachian and Ouachita mountain system that were formed 300 million years ago. Near Marathon, the range makes a surface appearance revealing even older rocks that were deformed in the process. Another smaller exposed part of the range is down inside Big Bend known as Solitario.

Big Bend

Big Bend aptly describes the course of the Rio Grande over a couple hundred miles of stretch. First the river bends left toward the northeast and then it gradually bends right as it resumes a southeasterly flow. The initial bend is situated between dramatic canyons: Santa Elena (Helena), Mariscal, and Boquillas. Luckily for rafters the canyons are downstream from the confluence of the Rio Conchos, which is upstream from Chihuahua and Coahuila political border, because it provides most of the surface flow. Big Bend National Park was established in 1944, so the park service manages environmental protection and recreational opportunities of a large area. Numerous geological and archaeological sites are found inside the park. The Chisos Mountains at the southernmost point are often shown on maps.

Upper Valley

As the Rio Grande enters Texas from New Mexico it flows through a relatively wide floodplain. At downtown El Paso, the American side is somewhat narrow because of the Franklin Mountains, but the right bank, with the sprawling city of Juarez, is not as constrained. Downstream on both sides for approximately 75 miles from El Paso/Juarez, the alluvial soiled plain is agriculturally valuable land with cotton and chili peppers. On the Mexican side, Maquiladoras dot the landscape away from the river. In Texas history, the Spanish founded numerous missions in the Upper Valley including the oldest continually settled place: Ysleta. East of El Paso, the all important Hueco Bolson field is situated between the Franklin and Hueco Mountains. The very deep bolson is capable of storing water like an aquifer, and the city utilizes this resource.

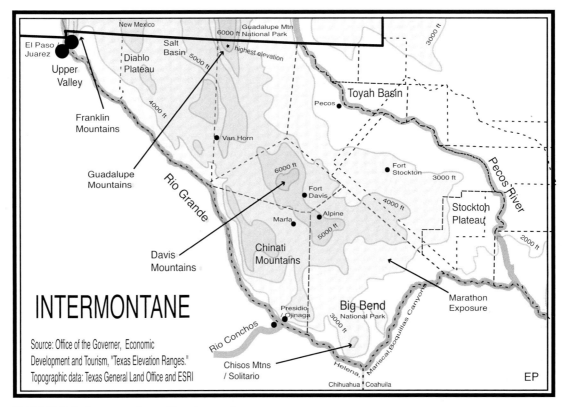

FIGURE 7-3 Intermontane Physiography

GREAT PLAINS

The Great Plains are quite extensive in Texas stretching from the northern Panhandle to the Mexican border. Much of West Texas, most of the Panhandle, and parts of Central Texas are all part of this North American physical province. Because of its size, it is often cartographically depicted as two parts: the High Plains and Edwards Plateau. Major places in the High Plains include Amarillo, Lubbock, and Midland-Odessa; Edwards Plateau has numerous smaller places like Fredericksburg. As mentioned earlier, the western boundary of the Great Plains is normally considered the Rocky Mountains, which in New Mexico works well enough. However, as the surface expression of the Rockies diminish, it is necessary to think of the Great Plains and Intermontane actually bordering one another. For convenience, the western edges of the Toyah Basin and Stockton Plateau are drawn as the boundary. It is possible to correlate the following political boundaries with the Great Plains in Texas: the northern and western Panhandle border segments and the Rio Grande upstream from Del Rio. The eastern boundary of the Great Plains has two dramatic and cartographically distinct features. The first is the Caprock Escarpment that separates the High Plains with lower elevated plains of the Interior Lowland. The second is the Balconies Escarpment that separates the Edwards Plateau with the broad Coastal Plain.

High Plains

The High Plains of Texas appears to be extremely flat as one experiences it at the surface. It may even be the most level section of the Great Plains. Yet, there is a distinct slope and aspect to the High Plains; basically the northwest is higher than the southeast. Therefore, drainage does trend toward the east, southeast, and south. The elevation in Dallam County in the northwest corner of the Panhandle is

FIGURE 7-4 Guadalupe Peak and El Capitan
South end of Guadalupe Range. Photo by Author.

FIGURE 7-5 Davis Mountains, Jeff Davis County
Mountain ecosystems along State Highway 17. Photo courtesy of Texas Department of Transportation.

FIGURE 7-6 Hueco Tanks, El Paso County
Hueco Tanks State Historic Site. Photo courtesy of Texas Department of Transportation.

approximately 4,700 feet. In Garza County at the edge of the Caprock, the elevation is approximately 3,000 feet. As both elevations suggest, the modifier "high" is relative to the even lower elevations to the southeast and not a statement of upland or mountainous terrain. While the Brazos and Colorado Rivers both currently originate on the High Plains, the relatively flat terrain generally produces incoherent drainage. Large and small areas are inundated with runoff creating various sized playas usually called playa lakes and sometimes buffalo wallows.

Canadian Breaks

The Canadian River slices across the Panhandle from its source in the Rockies to its eventual merger with the Arkansas River. The river has created a floodplain with distinct edges along its banks. Early cattlemen considered the crossing treacherous because of the quicksand of the floodplain and it came to be known as the Canadian Breaks. The Breaks are also a convenient line to separate the central and southern sections of the High Plains. The area *"North of the Breaks"* is classified as the Central High Plains. The area *"South of the Breaks"* is then classified as the Southern High Plains. The High Plains and more properly the Southern High Plains is known by a popular Spanish derived name: *"Llano Estacado"* (pronounced yano estacado). The Llano Estacado is usually translated into English as the "Staked Plain" and commonly attributed to the Coronado Expedition.

Toyah Basin and Stockton Plateau

Both the Toyah Basin and Stockton Plateau serve as the western boundary of the Great Plains in Texas. As mentioned previously, many generalize the westernmost part of state as the Trans-Pecos that implies the Pecos River is a physical boundary. The lower stretches of the Pecos in Texas does provide a stark line because the river is entrenched between the Edwards and Stockton Plateaus. However, the two plateaus are structurally similar, and the river has maintained its course as uplifting occurred around it. Because of the general trend, the Stockton Plateau has higher surface elevations. The Toyah Basin is an open expanse of flat land that was once part of an ancient sea bed. Currently, water from the Pecos River is used to irrigate crops along its course, so the Toyah Basin has relative agricultural value, for example, Pecos' melons.

Edwards Plateau

The Edwards Plateau differs from the rest of the Great Plains in two ways. Firstly, the plateau is lacking alluvial cover, which is why some scholars consider it to be an example of a stripped plain. Over geologic time, the Pecos River has captured the headwaters of the Colorado, Brazos, and Red Rivers sequentially. Without any current source of sediment (alluvium) from the mountains to the west, the surface material has eroded away. Other scholars question whether the Edwards Plateau ever had an alluvial surface. Secondly, the surface is predominately limestone today. Because most of the plateau is climatically arid, the limestone surface is relatively hard, which makes water infiltration difficult. During precipitation events, the water must move laterally and sometimes causes flash flooding. The water that does infiltrate

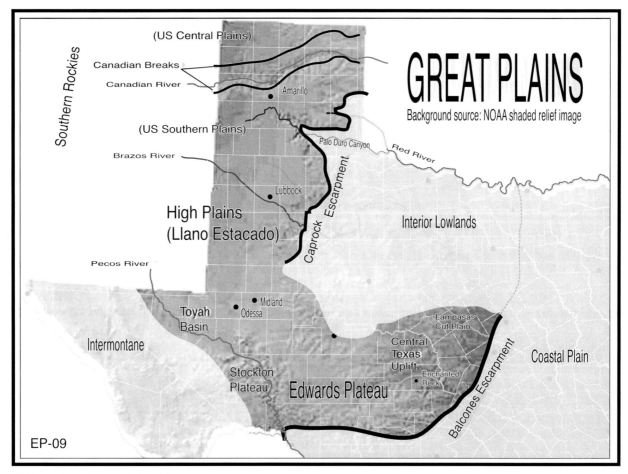

FIGURE 7-7 Great Plains Physiography

FIGURE 7-8 The Lighthouse, Palo Duro Canyon
Photo courtesy of Texas Department of Transportation. (Previously published in the State Travel Guide.)

acts different in the limestone as it actually chemically dissolves the limestone. In effect, it erodes from beneath. After enough time, caverns and canyons develop. This is very prominent along the southeast edge of the plateau as it comes into contact with the Coastal Plain.

Canyonlands

The boundary between the Edwards Plateau and Blackland Prairie is geologically distinct at the Balcones Fault Zone (BFZ). While not being tectonically active today, there is a general change of elevation associated with the fault zone. About 10 million years ago the Edwards Plateau was uplifted about 2,000 feet, and the fault zone indicates the approximate boundary of this uplift. *Balcones Escarpment* is the term applied to the boundary where enough topographic relief currently exists—roughly between Del Rio and Temple. The maximum elevation relief occurs near Del Rio with about a 1,000-feet difference; in Austin the escarpment is approximately 400 feet; at Waco the boundary becomes indiscernible. Along the escarpment, erosion from springs and rivers produces a variable terrain. Recently, ecologists are classifying this small area as the *Balcones Canyonlands*. Historically, the term *"Hill Country"* was applied to the same general area. From the coastal perspective, the escarpment relief and canyon topography appears to be hilly, which may have been a welcome contrast for early settlers on the coastal prairies.

Lampasas Cut Plain

The easternmost part of the Edwards Plateau is distinct because the limestone surface is exposed to more precipitation and humidity. Topographically, this is called a cut plain when the limestone becomes "sculptured" by the extra water. Despite the continuity with the Edwards Plateau, the Lampasas Cut Plain is sometimes classified with the areas to its north such as the Cross Timbers and Grand Prairie.

FIGURE 7-9 Plains Covered in Snow
Lake Meredith National Recreation Area, Moore County. Photo courtesy of Texas Department of Transportation. (Previously published in *Texas Highways*.)

Central Texas Uplift

The Central Texas Uplift goes by a variety of other names but there is no arguing about the physical distinction. In fact it is so distinct that it really doesn't fit into any of the major physiographic provinces properly. By default, we generalize it into the discussion of the Edwards Plateau. Early explorers immediately recognized the variety of minerals and speculated that the area would become a mining region. The name *Central Mining Region* became popular despite the fact that no systematic mining developed. A second name often used is the *Llano Basin* which reflects the immediate topography at the surface along the Llano River. However, the geological structure is actually an uplift (and not a basin). Therefore, one sees a corrected *Llano Uplift* and an entirely new term, *Central Texas Uplift*. The uplift is also unique because the age of the granite and schiest materials is extremely old dating to the oldest geological category of Precambrian. After the Ouachita Range developed, these Precambrian rocks became covered by more recent sediments that are currently being eroded. Enchanted Rock is a "batholith" or protruding part of this old granite that became exposed when the surrounding materials eroded away.

FIGURE 7-10 US-90 Bridge over the Pecos River
The Pecos divides the Edwards Plateau and Stockton Plateau. Photo by Author.

INTERIOR LOWLANDS

Only a small fraction of the extensive Interior Lowlands is found in Texas. Generally speaking the elevations are lower than the Great Plains but slightly higher than the Coastal Plain. The western boundary with the High Plains is very distinct with the Caprock Escarpment. The headwaters of the Red and tributaries of the Brazos are eating away at the Caprock slowly expanding the lowlands. By default, the northern boundary is the political border with Oklahoma along the Red River. The eastern boundary is a subtle one through the Metroplex, and it is typically drawn along the Eastern Cross Timbers which coincides with the submerged Ouachita Range. The southern boundary is less exact cartographically because the surface features transition together and is open to interpretation. Some of the major places located in this physiographic province include Fort Worth, Wichita Falls, Abilene, and San Angelo.

Osage Plains

The Osage Plains cover nearly two-thirds of the Interior Lowlands. It is commonly called other names such as the North-Central Plains; in some instances the terms lower (elevation) and red (soil color) are used. In some areas the surface is extremely level that gives it the proper name of plain. However, there are numerous areas of rolling hills or gentle undulations of the surface that are definitely not flat. In the south, a remnant known as the *Callahan Divide* stands out as it separates the Brazos and Colorado Rivers' drainage basins.

West Texas Rolling Plains

The westernmost part of the Osage has more topographic variety with slightly higher elevations. The region typically goes by the name West Texas Rolling Plains reflecting that there is more topographic variation. Those areas with red-colored Permian derived soils are sometimes labeled red, such as the Red Beds and Red River Rolling Plain, yet other names such as Gypsum Plains can be found on certain maps.

152 Part Three Physical Geographies of Texas

FIGURE 7-11 Hill Country Overlooking Colorado River
Suburban sprawl extends upstream along reservoirs. Photo by Author.

FIGURE 7-12 Enchanted Rock, Central Texas Uplift
Enchanted Rock State Natural Area. Photo by Author.

Grand Prairie

The eastern third of the Interior Lowlands are often classified by only one of two very different ecosystems: Grand Prairie and Cross Timbers. The Grand Prairie is nestled between the Cross Timbers and extends from close to the Red River down to Waco in terms of vegetation. Occasionally, it is referred to as the Fort Worth Prairie, and the city of Grand Prairie is on the eastern edge. Underlying the Grand Prairie is the *Comanche Plateau,* which on some maps extends southward underlying the Lampasas Cut Plain. Regardless of the cartography, the Grand Prairie does have a limestone base that influences the natural vegetation. It is a grassland instead of a woodland despite receiving enough precipitation to support a forest environment similar to the Cross Timbers.

Cross Timbers

The Cross Timbers are the woodland environments that separate the grassland plains/prairies in North Central Texas. The sequence is as follows: Osage Plains, *Western Cross Timbers,* Grand Prairie, *Eastern Cross Timbers,* and Black Prairie. Technically, the two Cross Timbers connect together along the Red River, and the vegetation type extends northward as part of the Central U.S. Hardwoods. Oaks are the

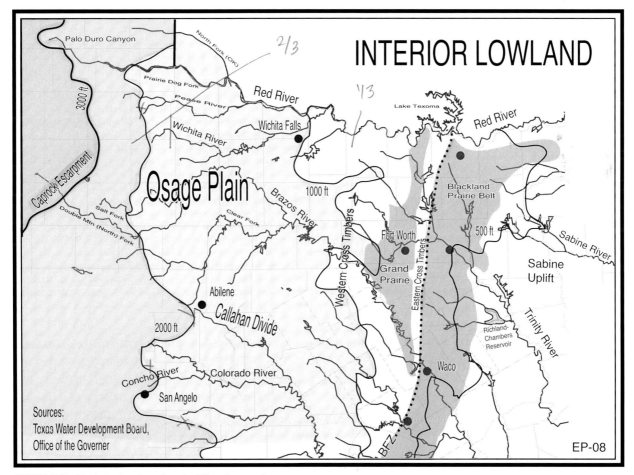

FIGURE 7-13 Interior Lowlands Physiography

dominant tree species but the understory contained numerous shrub species that made early travel difficult. The Western Cross Timbers are more expansive and historically the Upper Southerners that arrived here transplanted cultural traits similar to the Ozarks and Appalachia. Underlying parts of the Western Cross Timbers is the *Palo Pinto* which is a geological remnant of sorts. The Palo Pinto has carbonate material instead of limestone and is exposed due to the erosion of sediments westward. The Eastern Cross Timbers are a less expansive line of woodlands that run in a north–south direction separating the Black and Grand Prairies. It turns out to follow the subsurface divide of the Ouachita Range. While the transition between the Interior Lowlands and Coastal Plain is very subtle, most but not all maps will use the Eastern Cross Timbers as the border.

GULF COASTAL PLAIN

The coastal plain that we find in Texas is part of a larger Atlantic Ocean coastal plain that extends beyond our immediate Gulf of Mexico coastline. For convenience and regard to geographic scale, I'll call this province the Gulf Coastal Plain instead of the longer Atlantic/Gulf Coastal Plain. The Gulf Coastal Plain extends hundreds of miles inland and occupies slightly more

FIGURE 7-14 Osage Plains near Red River Border
FM 677 near Illinois Bend, Montage County. Photo courtesy of Texas Department of Transportation.

surface area than the Great Plains in Texas. Locating the coastal plain is easy with it being immediately adjacent to the Gulf of Mexico. Because the plain extends beyond Texas, the political boundaries with Tamaulipas, Nuevo Leon, Louisiana, Arkansas, and the downstream sections of Coahuila and Oklahoma are the de facto border. The inland border follows the geological "S" line along the historic Ouachita Range, which is also the Balcones Fault Zone. Along the exposed Balcones Escarpment, the border with the Great Plains is distinct. Yet the border with the Interior Lowlands is more transitional in nature with the Eastern Cross Timbers serving as the delineator. The Gulf Coastal Plain of Texas consists of extremely thick layers of sediment that were transported and deposited by the rivers originating northwest of the "S" line. During various stages, the coastline moved from the "S" line farther southeast with the previous coastal landforms left behind. The general pattern as expected reveals younger surfaces are closer to the present coastline, and because these processes continue to occur the actual coastline is constantly changing. Much of Texas history played out on the coastal plain and its economic importance to the state cannot be overstated. Foremost, most Texans live on the Gulf Coastal Plain. Likewise the four largest cities, Houston, San Antonio, Dallas and Austin, are located on the coastal plain as are Texarkana and Laredo on the borders.

Blackland Prairies

The Blackland Prairies are the oldest of the three broad concentric regions of the Gulf Coastal Plain. Being farthest inland, it logically borders the other North American physiographic provinces. The generic label "prairie" comes from the vegetation which included both short and medium grasses. The specific name, "blackland," refers to the soil color, which was noticed by early American settlers and considered to be relatively fertile. The blackness of the soil comes from the high organic content that most grasslands have because of the in-situ decay of previous growth. The fertility of these soils permits intensive mechanized agriculture that we see today except that some of this land is also coveted for suburban expansion. Many maps label the region as the "Blackland Belt," which aptly describes the pattern. The Blacklands extend from the Red River at its wider extent to the Rio Grande at Del Rio. Many cartographers label the section near Dallas more exactly with the place-name "Black Prairie." When mapping the prairie ecosystem, a few outlying prairies exist; they are sometimes labeled with their own names such as "Fayette Prairie" and "San Antonio Prairie."

FIGURE 7-15 Road through Cross Timbers Vegetation
Mixed oak woodland environment, Callahan County. Photo by Author.

Interior Coastal Plain

The Interior Coastal Plain is sandwiched between the Blackland Prairies and Coastal Prairies. It is also intermediate in age between the more inland and more coastal subregions. Two large areas, East Texas and South Texas, with very different ecosystems anchor the Interior Coastal Plain with a large third area that links them together.

East Texas Timberlands

The eastern part of the state is dominated by coniferous forest with numerous pine species. The East Texas Timberlands extend from the Arkansas border to about 25 miles from the Gulf of Mexico and extends westward about 75 to 125 miles. Loblolly pine is the most common tree not only in East Texas but probably the whole state. Long leaf pines were more common before widespread logging yet the loblolly was the preferred replanting tree. Hardwood trees such as hickory are also common especially around rivers and creeks. Regardless of exact species, the region has been identified with the pine; the colloquial region name

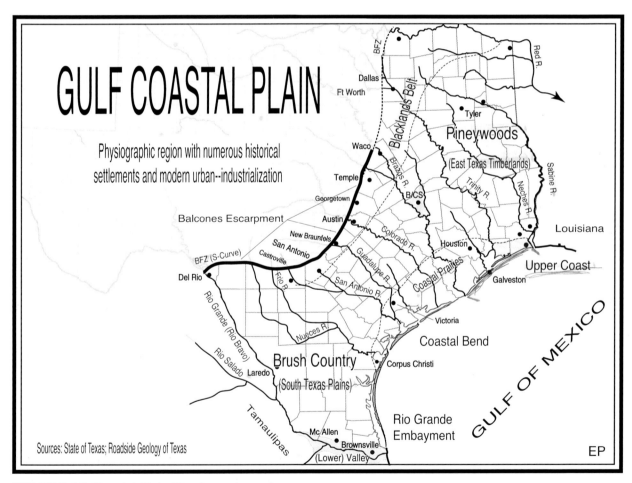

FIGURE 7-16 Coastal Plain Physiography

is "Piney Woods." Both coniferous forest and the term Piney Woods are found east of Texas all the way to Georgia. The timberlands of East Texas and the American Southeast support commercially viable forestry. Near the headwaters of the Sabine and Neches Rivers, a slightly higher elevated area aptly called the *Sabine Uplift* often appears on maps. Likewise, the same area is denoted on other maps as the *Sandy Hills*. Farther south, an area of the timberlands that was difficult to settle was "Big Thicket." Big Thicket is often described as an ecological crossroads, and it has been part of the National Park System since 1974 as a "National Preserve."

Post Oak Savannas

The wooded environment west of the Piney Woods is dominated by hardwood oaks. Instead of a densely forested region which prevents direct sunlight at the surface, the landscape reveals more of a savanna environment with trees and grasses coexisting. The post oak is the most common tree species and the region is named after it, however other oaks and trees are common. A micro-geography of vegetation exists between upland/sandy and lowland/clay environments and it is very observable near rivers and creeks. The Post Oak Savanna can be conceptualized as a transitional or continuum between the pine forests and grasslands. On many maps the region is labeled as the "Post Oak Belt" reflecting the linear pattern from northeastern Texas southwestwardly toward San Antonio. Similar to the Blackland Prairies, there are smaller outlier regions interspersed with the prairie outliers.

South Texas Plains

Much of South Texas falls onto the Interior Coastal Plain. In addition to the label South Texas Plains, *Nueces Plain* might be on a map in the Nueces River drainage basin. Other labels such as "Brush Country" are often used interchangeably. Technically, most of the South Texas Plains is wooded (chaparral), but the growth is considered stunted. The combination of warmer and drier weather affects not only vegetation but also drainage patterns. Along the coast between the Nueces River and Rio Grande, there exists a good example known as the "South Texas Sand Sheet." The Sand Sheet actually derives its sand from the beaches along the Gulf, but no established rivers transport it away before it can harden and stabilize.

Coastal Prairies

The Coastal Prairies parallel the Gulf of Mexico coastline reaching inland 25 to 60 miles. This is the most geologically recent part of the coastal plain, and the flat topography, wet soils, poor drainage, and hygrophyte vegetation reflect this newness. Some maps show the Coastal Prairies as running from the Sabine River to the Rio Grande, however, others show a smaller, more constrained area where prairie grasses actually exist. The difference concerns the cartographers' intent. The larger is based on the geological factors and the smaller interpretation is vegetation. The quality of the grasses in the proper coastal prairie vegetation supports the most cattle per area in Texas. At the coastline, there are a variety of coastal landforms such as barrier islands, estuaries, and sand dunes. The remaining discussion of the Gulf Coastal Plain follows the coastline in sections from north to south.

Upper Coast

The Upper Coast of Texas begins at Sabine Lake and Pass and extends down to the Brazos River. The Upper Coast near the Louisiana border has a narrow section of coastal prairie between the Bolivar Coast and the Big Thicket. However, the immediate coastal area is very marshy and that confined any possible east-west corridor. The prairie gradually becomes wider as one moves westward. The Upper Coast receives the highest average precipitation amounts in the state. The famous oil gusher at Spindletop (outside Beaumont) is located on the Upper Coast that propelled numerous developments especially toward the oil and gas business. In fact, the region is highly affected by the dual industrialization and urbanization processes which set it apart from the rest of the Texas coast. Houston sits in

FIGURE 7-17 Blackland Prairie, Milam County
Land being used for intensive agriculture. Photo by Author.

the middle of this economic and demographic growth, and the three places of Beaumont, Port Arthur, and Orange anchor the southeast.

Galveston

The dramatic topographic feature of the Upper Coast is *Galveston Bay*. As a natural feature, Galveston Bay is the terminus for the Trinity and San Jacinto Rivers, and then the primary opening to the Gulf is in between *Galveston Island* and the Bolivar Peninsula. On the northern bayside of Galveston Island is the city of Galveston. The northeast lobe of the bay is often labeled as Trinity Bay. At the northwest end of the bay is Buffalo Bayou where the city of Houston was founded along.

Coastal Bend

The Coastal Bend literally refers to the general shape of the Texas/Gulf of Mexico coastline. More specifically, it refers to the area between Corpus Christi and the Brazos River. While the Coastal Bend has similar petroleum resources as the Upper Coast, the industrial footprint has been much smaller. Relatively natural areas along the coastline have been preserved, and small towns including fishing villages are located at various points. A series of barrier islands and estuaries line the Texas coastline.

Matagorda

The first or northernmost is *Matagorda Bay* and *Matagorda Island.* At the north end of the bay, the Colorado River terminates, and the Lavaca River empties on the eastern margin where Port Lavaca is today. La Salle's failed colony Fort St. Louis and Galveston's early seaport and immigration competitor, Indianola, were on Matagorda Bay.

FIGURE 7-18 Piney Woods, Houston County
Mission Tejas State Historic Site. Photo by Author.

San Antonio and Aransas Bays

San Antonio Bay is the terminus for the San Antonio and Guadalupe Rivers. Immediately south of San Antonio Bay is Aransas Bay. Nestled between the bays is a previous coastline that has been subsequently shielded by newer barrier islands; Aransas National Wildlife Refuge is located here with its incredible bird species diversity. Located on the backside of Aransas Bay's barrier islands is one of the most complete wash over marsh systems in Texas. The muddy tidal plain that seabirds love is covered during high tide and is exposed at low tide.

Corpus Christi Bay

Corpus Christi Bay is formed by the inundated mouth of the Nueces River with a barrier island (Mustang Island) protecting it from the open sea. The city of Corpus Christi was founded on the south side of the bay. Numerous military facilities are located in and around the city since it is the last natural port for naval operations along the U.S. coastline. Access to the northern end of *Padre Island* where the National Seashore Park is located is through Corpus Christi.

FIGURE 7-19 Post Oak Savanna, Fayette County
Cattle in bluebonnet field. Photo courtesy of Texas Department of Transportation.

FIGURE 7-20 Amistad Reservoir, South Texas Plains
Texas' second largest reservoir is shared with Mexico. Photo courtesy of Texas Department of Transportation.

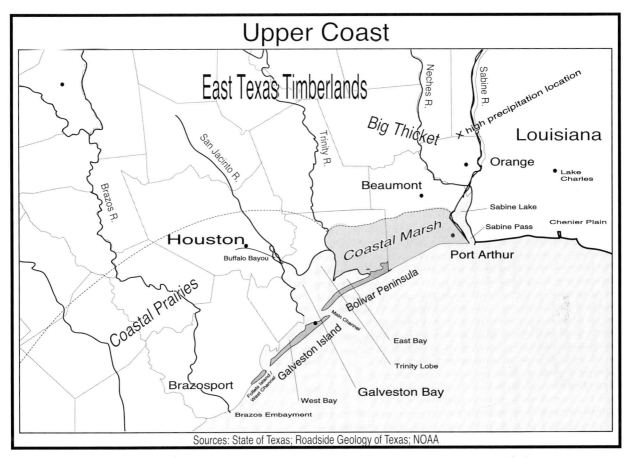

FIGURE 7-21 Upper Coast of Texas

FIGURE 7-22 Coastal marsh, Indianola
Salt grasses flourish along the tidal margin. Photo by Author.

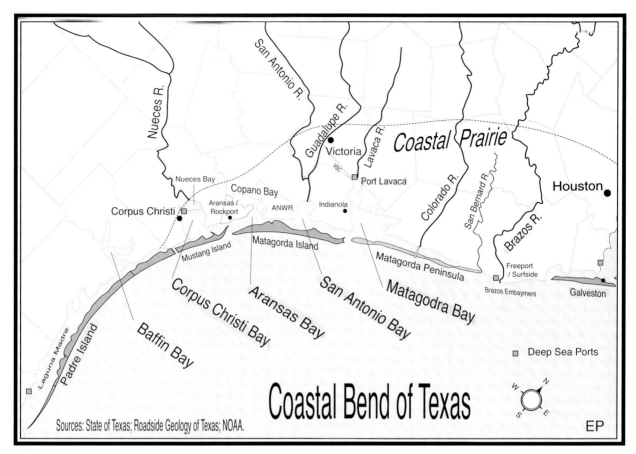

FIGURE 7-23 Coastal Bend of Texas

Rio Grande Embayment

The final segment of the Texas coastline is the Embayment. Along the coast, a similar process of barrier islands and estuaries continue southward. The difference is that the Rio Grande deposited extremely large amounts of sediment in geological time to keep up with the rising sea levels. So the coastline gently bends seaward around the mouth of the river which helps create the southern cartographic "bend" of the Coastal Bend. Despite lots of material, it isn't enough to form a delta like the Mississippi River. The river's sediment provides material for the adjacent barrier island complex. While Brazos Island is the southernmost, Padre Island is the longest barrier island in the system stretching all the way to Corpus Christi. The southern tip of Padre Island *(South Padre Island)* has causeway access to the mainland and has become a major tourist and recreational place. The backside of Padre Island goes by the name Laguna Madre, which is very shallow except where the Intracoastal Waterway was dredged. *Baffin Bay* is another major estuary except it has no significant river terminus associated with it. There is a channel across Padre Island near the entrance to Baffin Bay, so some maps label the island(s) North and South respectively. On the landward side of Laguna Madre is the South Texas Sand Sheet, which forms from sand that is transported by air from the barrier islands.

Lower Valley

Simply called "The Valley" by many Texans, it is as much a physical region as it is a cultural one. Even the term *Rio Grande Valley* is not specific enough because it doesn't distinguish between the upper and lower valleys. The Lower Valley is an alluvial floodplain (not really a valley) created by the Rio Grande stretching from Rio Grande City to the Gulf of Mexico. With good soils and a warm climate, the Lower Valley requires reliable water to be agriculturally valuable. Two major reservoirs were built on the Rio Grande to insure water delivery, and the U.S.–Mexico Border Commission manages both. The best farmland is on the immediate floodplain as well as any adjacent lands with access to irrigation water systems. Recent years have seen major urban development in the Lower Valley.

GULF OF MEXICO

The Gulf of Mexico is a significant physical feature in its own right. For Texas, it is a major factor in our climate and it is one of the longer political border segments. The Gulf of Mexico has its own geography with numerous physical characteristics such as location, temperature, and three-dimensional shape. At a macro-geographical scale, the Gulf of Mexico is the ninth largest body of water on Earth. The outline of the Gulf follows a roughly circular pattern with two "openings" that directly link it with waters of the Atlantic Ocean. The *Straits of Florida* separate the Florida Peninsula with the island of Cuba; the *Yucatan Channel* separates the Yucatan Peninsula with the island of Cuba and technically enters the Caribbean Sea. Along the southwest margins of the Gulf, most maps will label the *Bay of Campeche* (or *Bahia de Campeche* in Spanish).

One defining feature of the Gulf is the "enclosure" of its shape; oceanographers would call this an enclosed sea. Enclosure means that the waters do not intermix quickly with the open oceans, so they can develop slightly different temperature and chemical properties. The Mediterranean Sea is the classic enclosed sea where recent research has shown how pollution has trouble dissipating along the industrialized coastlines. It's been common knowledge for a long time that the Mediterranean's moisture and temperature influences

FIGURE 7-24 Boat dry docks, Matagorda County
Shipping and fishing infrastructure along the Gulf of Mexico.

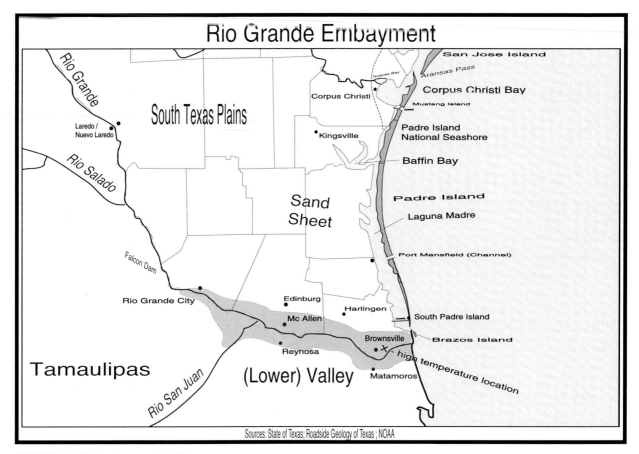

FIGURE 7-25 Embayment of Texas

climates in Southern Europe and North Africa. The waters of the Gulf are situated at sub-tropical latitudes between 18 and 31 degrees N, which means it is potentially warmer than the Mediterranean. Without a doubt, the Gulf is a warm body of water. In winter, it mollifies colder air, and along the coast it is very rare for a below freezing temperature. It also moderates temperatures in summer, but the humidity sort of masks the difference. However, warm water is a reservoir of energy, hurricanes have the potential to strengthen and not weaken as they move out of the tropics and into the Gulf.

The Gulf of Mexico has a morphology that resembles the Atlantic Ocean but on a smaller scale. A profile look at the Gulf reveals that it has both shallow shelf, deep oceanic crust, and a slope between them. In geological time, the Gulf opened up during plate tectonic movements creating a deep floor over 3,000 meters deep. This deep oceanic floor is so flat it is called a plain, specifically an abyssal plain. In the Gulf the name "Sigsbee" is used: *Sigsbee Abyssal Plain.* Along the current shoreline, the Gulf is relatively shallow. This shelf drops only 200 meters per 100 miles or more offshore. This shallow shelf helps the offshore oil drilling community tremendously. The width of the shelf varies, and three large areas of shelf are given their own specific names: *Texas-Louisiana Shelf, West Florida Shelf,* and *Yucatan Shelf.* In contrast to the other two that have relatively steep escarpments, the Texas-Louisiana continental margin has a more gradual slope. East of the *Texas-Louisiana Slope* there is a large alluvial fan associated with the Mississippi River. The *Mississippi Fan* accumulates enormous amounts of sediment that deposit not only on the shelf but out over the slope. An interesting feature about 100 miles from Galveston is the *Flower Gardens.* The Flower Gardens consist as two biologically rich reefs (named East and West) that sit atop of bulging salt domes. Currently, the reefs are protected and managed by the Flower Garden Bank National Marine Monument.

Chapter 7 Physiographic Regions 165

FIGURE 7-26 Barrier Island Development, South Padre
South Padre Island, Cameron County. Photo courtesy of Texas Department of Transportation.

FIGURE 7-27 National Seashore, north Padre island
North end of Padre island differs much from the south end development. Photo by Author.

166 Part Three Physical Geographies of Texas

FIGURE 7-28 Gulf of Mexico

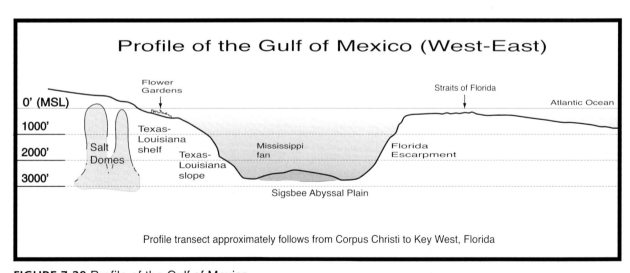

FIGURE 7-29 Profile of the Gulf of Mexico
Source: E. Swanson, 1995. Geo-Texas (p.147). The vertical cross-section through the Gulf reveals both deep and shallow waters. The shallow Texas—Louisiana shelf is an extension of the Coastal Plain except it is currently submerged. The deepest waters of the Gulf are over 3000 feet which is similar to deep parts of the oceans

Circulation of water in the Gulf of Mexico can be determined and mapped. Because the Gulf is enclosed, the overall circulation is impacted by where the openings are located. Water in the eastern Gulf circulates more than the water in the west. The majority water entering the Gulf comes in through the Yucatan Channel. Much of this water flows northward and quickly exits eastward through the Straits of Florida. This quick path in the eastern Gulf is known as the Loop Current. This warm water moves northeastward along the East Coast of the United States, and it is called the Gulf Stream. As part of the general North Atlantic clockwise circulation, warm Gulf Stream water eventually impacts the weather of Northern Europe. In the western Gulf, a clockwise moving gyre is a constant feature. Along the Texas coast, water moves slowly alongshore but the direction is seasonal. The long-shore current in summer is northward and in winter it is southward. Tidal surges are another way to circulate seawater; however, tides in the Gulf are minimal. Except during storms, the tidal surge only averages less than half a meter.

The Gulf of Mexico is an economical and ecological asset for both the United States and Mexico. Because of the different territorial water definition, Texas' share of the Gulf is potentially greater than other states; the real political factor is the exact locations of fossil fuels. As an enclosed sea, the Gulf of Mexico shares similar ecological concerns as the Mediterranean Sea. Like the Mediterranean, numerous political entities share the resource with different interpretations about environmental considerations. In the United States, places closer to the coast are growing faster than the country as a whole. Over one-third of all Texans live within a hundred miles of the Gulf, and one-third of the economic activities are located in the same range. Therefore, a third of the state in terms of people and wealth are in the potential path of a major hurricane. Considering the population and business dimensions near the Gulf, Texans need to understand our relationship with this physical region.

In this chapter, physical regions were presented as a hierarchy of location. Each place has its own unique physical story, but the regions they fall in are useful generalizations for this discussion. With the 50 or so mentioned regions, a picture of the physical diversity of Texas becomes apparent. In the next chapter, we turn to the human–environmental relationship to explore how Texans interact with the physical geography around them.

FIGURE 7-30 Corpus Christi Waterfront
Integrated seawall structure with pedestrian uses. Photo by Author.

"Human Agency Impacts the Environment"

FIGURE 8-1 Riverside Development Along Colorads River
Residences and power lines near Granite Falls, Burnet County. Photo by Author.

8

Human–Environment Interactions

The environment provides us humans with the materials necessary to live. Humans have an essential and somewhat commandeering relationship with the environment. Basic manipulations of the environment to create better living conditions and better habitats for those plants and animals we covet have consequences. As if there needs to be another reason to study the environment, scholars are questioning the fundamental nature of the relationship because of undesirable results. As we've become socially aware of global warming, the idea of extreme consequences to humans is being discussed in a more open manner.

In this chapter, we discuss the human–environment relationship. We begin with the idea of humans as agents of change. Secondly, we discuss agriculture and biodiversity with some level of detail concerning Texas agriculture. Then, we discuss the environment through the lens of resources and hazards. Finally, we conclude with some ruminations about environmentalism in Texas.

HUMANS AS AGENTS OF ENVIRONMENTAL CHANGE

As we saw in previous chapters, human agency is a potent force. Human mobility and technological change have driven political and economic processes for hundreds of years—shaping things like the historical geography of Texas. Human agency also plays an important role in the physical world. Many people prefer to separate culture from nature, which extends into separate human geographies and physical geographies. An expansive view of culture states that it shapes society, and even the most mundane artifacts like freeways and shopping centers have strong cultural meaning. Nature is supposed to have its own rules that shape or control things like forests, wildlife, and sedimentation. Although it might be nice to believe in a romantic nature, the opposite is true. Humans not only construct nature as an idea in our minds, we directly influence the physical world around us. In many ways, human's direct and indirect actions shape the forests and wetlands.

One way to think about these modifications is not to separate people from their surroundings. We are a part of the environment, and we affect it by our mere presence. Think about any school or college campus, if you look attentively you'll probably notice where students take short-cuts across grassy areas. Along frequently crossed areas, the soil becomes compacted by all the footsteps that strike the ground and grass has difficulty growing. Just like other animals, we create pathways that are physically different from the immediately adjacent areas. Unlike other species, we can contemplate the meaning of a pathway and the path less traveled. Quite literally we do leave a footprint as individuals, and collectively we create potentially huge footprints.

Historically, human impacts were minimal but they did occur, and the impacts become greater with technology. Early humans changed their environments through selection strategies. By choosing a favorite plant or animal, we changed those species. Picking fruit from the tastier or more nutritious plant, we spread those seeds as we wandered. Later, as we identified the preferred plant, we eliminated the competition to maximize the potential crop. In terms of animals, we hunted some species to extinction, which isn't a good strategy per se. Yet it made way for a larger number of other species that we eventually domesticated. Domesticated animals and plants are exposed to further selectivity and manipulation as humans begin to control the reproduction of these species. Finally, humans modify the environment through tools and technology associated with selection strategies. Historically, the use of fire has been important to humans because of how it betters our lives. Fire in the form of a campfire goes beyond the altering of our food before consumption and it has many connotations such as the safety, warmth, and companionship of hearth. The collection of firewood and fire starting items become another selection factor of human agency. Another is the widespread burning of grasslands to enhance foraging for prey which inhibits tree seedlings and forest expansion. Prehistoric peoples routinely burned select areas for these long term benefits. Fire by natural or human origins changes the composition of the vegetation toward those species that can survive routine burning and even thrive in the new bio-chemical environment afterward.

Contemporary people alter the environment even if they don't have an immediate relationship to agriculture and campfires. Developed societies rely on the sophisticated delivery of food and energy from other locations that still have that direct environmental interaction. The city is the antithesis of rural agriculture, and many of our good examples of anthropogenic change come from cities. The close concentration of buildings in a city creates a different surface for sunlight, rainfall, and runoff. Urban heat island is a long recognized phenomenon that cities constantly register warmer high temperatures and larger temperature ranges than the surrounding rural areas. The predominance of concrete, brick, mortar, and asphalt changes the heat absorption of solar radiation when compared to vegetation. Likewise, when it rains, raindrops strike roofs and parking lots that do not absorb moisture and direct the water elsewhere. Cities as human artifacts alter the immediate climatic and hydrological processes. Furthermore, widespread leveling and shaping of the surface as well as vegetation removal is common for residential and commercial development. Even suburbia because of its large scale requires complex infrastructure such as utilities, highways, and flood control.

Agriculture

Modern agriculture has many nuanced geographies that reflect human modification to the environment. In addition to preferential selection of species and most ideal sites, agriculture has adopted technological changes. With the advent of mechanized agriculture, fewer people need to be active workers and the equipment can plow, plant, and harvest the crops. Technologies that pump water from aquifers may allow irrigated crops to be located in relatively arid locations. Because of the combination of large size, land use decisions, and the utilization of technology, Texas is number one in many agricultural statistics. Texas has more than double the number of farms and twice the amount of farm acreage than the second ranked state. An amazingly high 75 percent of all land in Texas is classified as agricultural, and agriculture is valued at close to 15 billion dollars per year.

Agriculture in Texas has evolved in the last 500 years with the arrival of migrants and diffusion of ideas as well as local innovation and adaptation. Native Texans practiced agriculture in the eastern side of the state when Spanish missionaries arrived. The initial European/American settlers to Texas had modest amounts of technology with a few metal tools. Most farmers though had experience with a variety of crops and animals that would be successful here in Texas. While the planters arrived with the goal of seeding cotton, upland farmers introduced mixed subsistence style of farming. Today the trend is to produce marketable products and follow increasingly corporate business models. Some of the larger agricultural activities are surveyed below.

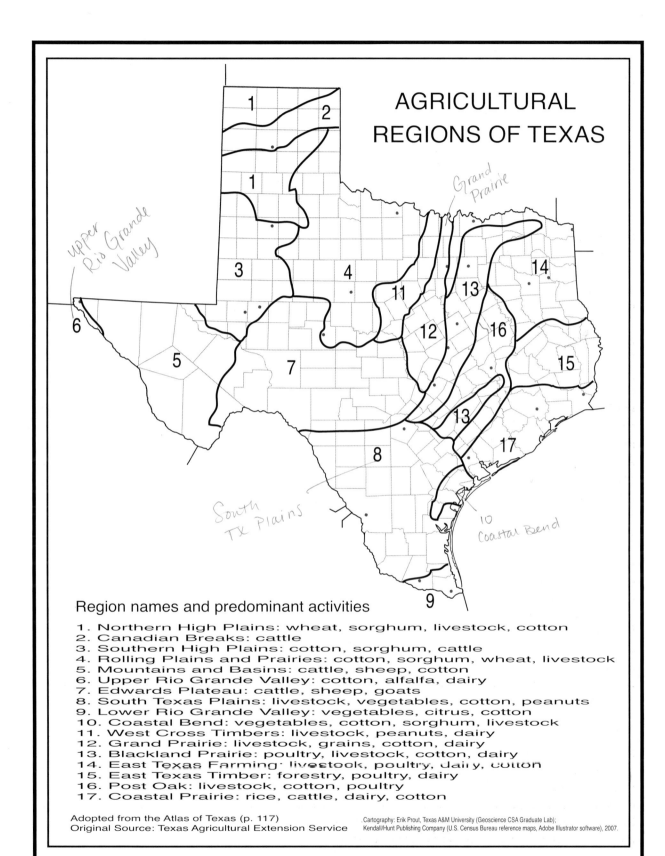

FIGURE 8-2 Agricultural Regions

TABLE 8-1 Agricultural Data

Products	Acreage/(inventory)	Value ($)
Cattle	(14,100,000)	8,083,024,000
Poultry	(97,400,000)	1,260,951,000
Sheep, Goats, Hogs	(3,200,000)	322,395,000
Total Livestock/Rangeland	**95,745,000**	**10,402,993,000**
Cotton	4,657,029	1,088,675,000
All forage	4,982,165	730,165,000
Grains	9,600,000	1,099,460,000
Nursery		704,699,000
Total Crops	**26,938,000**	**3,731,751,000**
Total Pastureland	15,914,000	
TOTALS	129,900,000	14,134,744,000

Sources: 2002 Census of Agriculture; Texas State Agriculture Overview—2005.
National Agricultural Statistic Service.

Cotton-Picking

Some of the same crops are still being grown today that were introduced by early planters. Cotton is one such crop. Initially introduced with plantations and slave labor, cotton was primarily located in East Texas and Gulf Coast along river bottoms. After the Civil War, cotton expanded inland and westward with the railroads; Dallas was situated well to become the financial center for cotton. Despite the boll weevil nearly destroying cotton production in the United States, it has bounced back with new types. The core area of cotton moved toward the Southern High Plains with the advent of aquifer-based irrigation. Between 5 to 6 billion acres are planted with cotton each year, and the value of the crop is around 1.5 billion dollars, which makes it the most valuable single crop in Texas.

FIGURE 8-3 Grain mill, Blacklands
Milam County. Photo by Author.

Chapter 8 Human–Environment Interactions 173

FIGURE 8-4 Livestock on the Panhandle
Photo courtesy of Texas Department of Transportation.

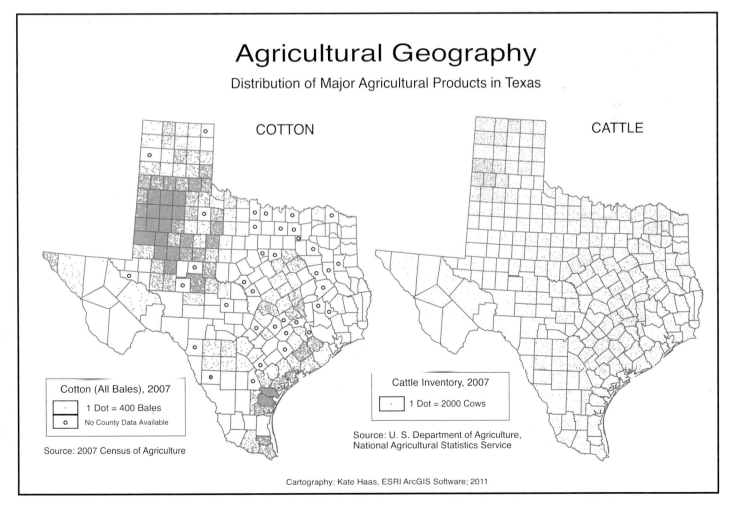

FIGURE 8-5 Agriculture: Cattle and Cotton

Other Crops

Excluding cotton, another 15 to 17 billion acres in Texas are planted with crops. With nearly the same total acreage as cotton, forage crops—principally hay—are collectively the second most valuable crop in Texas. Grain crops such as winter wheat, corn, and sorghum stand out for their acreage and/or commodity values. The nursery/greenhouse category also ranks fifth in terms of commodity value. Overall, the value of crops in Texas is estimated at 3.7 billion dollars, which ranks it in sixth place nationally.

Cattle-Driving

Livestock values exceed crops by quite a large margin, and it is driven by the large cattle sector. Total livestock values are nearly 10.5 billion dollars and that is a clear number one national ranking. At the top of the value list are cattle, and the traditions of cattle ranching dramatically impact the land use statistics. Over half of the state (95 million acres) is categorized as rangelands. Furthermore, the nearly 20 million acres classified as pastureland and planted with forage crops are primarily geared toward cattle. There are approximately 14 million cattle in Texas which amounts to only a 2:3 ratio with people. Cattle are found nearly everywhere in Texas and they are part of the most ubiquitous landscapes namely rangeland.

Other Livestock

Texas is also number one nationally in the number of goats. Both sheep and goats were introduced by the Spanish and are still found in areas where they are more appropriate than cattle; wool and mohair industries developed alongside the meat processing. Despite their importance to Central European sausage making, hogs and the resultant pork processing is quite small when compared to the national rankings. The rise of the poultry industry has elevated the chicken population past cattle by a wide margin. However, poultry tends to be an extremely intensive land use when compared to the extensive nature of cattle. Currently, broilers and layers are both ranked in the top 10 nationally. Without a doubt, livestock ranching has impacted the natural environment with so many hoofs (footprints) but also in other ways. Culturally, livestock ranching is one of the stereotyped images of Texas; ecologically, it is a testament to the biological changes of the last half a millennium.

Biodiversity

Agriculture changes the environment in many different ways, but one basic way is that humans are selecting and promoting specific species over others. The result of this selection process is a loss of biodiversity. Theoretically, natural processes evolve toward greater biodiversity with species and sub-species creating more nuanced niches for themselves while being spatially mixed together. In the likely situation of abrupt environmental disruption, some species will go extinct and some will survive and eventually diversify again. Most Americans have not seen a natural grassland; such a place if it exists would have a diverse, complex arrangement of different grasses in various stages of growth. Only subtle patterns of soil chemistry and soil moisture should be altering the concentrations of species. That same field today would have a single crop planted in neat rows or be seeded for better rangeland.

There are very different biodiversity rates around the planet. Tropical rainforests have the highest biodiversity rates in the world with correspondingly high amounts of carbon storage. Tropical savannas have the highest biomass amounts in terms of animal sustainability. Polar regions tend to have the lowest biodiversity rates with an almost identical environment circumscribing the Arctic in the northern hemisphere. The mid-latitudes with their alternating woodlands and grasslands should have biodiversity rates somewhere in between; however, the rise of modern agriculture especially "mono-culturing" has altered this zone tremendously. Texas is part of this "Neo-European" agricultural world dominated by livestock and grains remarkable similar to Northwest Europe as well as other places where European ideas and practices diffused such as Australia and Argentina.

Modern agriculture reduces biodiversity with two different processes. The first process is the simple displacement of space for non-selected species. As humans plant crops and tend animals, the literal space for their competitors is taken away. In a typical field, the soil is plowed to create ideal growing conditions and only a chosen seed type is planted. Weeds and pests are controlled as much as possible—typically with chemicals today. Logically, every acre planted is one less acre for other plants. Likewise, animals are bred and maintained in a space that is fenced to contain them and exclude other species. Both competitors and predators are kept to a minimum level and even eradicated in some instances. It has been argued that this component of mono-culture is very susceptible to problems. For example, a bug that consumes corn finds a paradise when it lands in the middle of a corn field and all the normal checks to infestation are gone. The concentration of similar animals as livestock becomes intensive means that infections spread quickly as well as a whole host of other sanitation problems associated with waste disposal.

The second process of biodiversity reduction is the inevitable selection of most desirable result. Since the beginning of agriculture, humans have propagated better sub-species through deliberate cross pollination and animal husbandry. The result is better crop harvest and more suited animals—most of the time meaning more and larger. Although high volume milk cows, large breasted chickens, and giant tasteless tomatoes are not implicitly a reduction of biodiversity, they point to a trend. With advanced technology and global scale of agriculture, farmers around the world can plant the identical high yield seed or genetically engineer the single best dairy cow. Currently, large percentages of the world's food production are coming from very few species/sub-species. This is most apparent in grain production with popular consumption of rice, wheat, and corn. There are hundreds of varieties of wild rice for example, but a fourth of the planet subsists on just a handful of mass produced types. Imagine the calamity of a crop failure based on just one type of rice or corn or wheat. Agriculture (humans) changes the environment. Many changes are beneficial to humans because they provide a better diet that allows us to do other creative and productive things. Other changes are fundamentally changing the composition of life on our planet, and they could potentially include negative feedbacks on humans themselves.

NATURAL RESOURCES

The environment as a broad concept is the material world of one's surroundings. As such, it has a variety of positive and negative "natural" elements that humans become aware of. The good elements are known as natural resources because they contribute to our quality of life. Contrarily, those elements that diminish our quality of life are known as natural hazards. Resources can be a direct good or an indirect benefit. In a direct and material manner, the earth that goes into making a brick or the vegetation that becomes part of a dwelling is a resource. Clean water from a stream and ample game for hunting are more examples of resources, which also implies the vantages of location. Indirect resources include energy because they have to be processed into the "good" that we recognize or take advantage of such as the light bulb.

Planning for the wise use of resources is one logical outcome of researching the natural environment. Humans have the absolute ability to plan and manage their resource use, but many societies often fail to do this well. The commoditization of resources presents a difficult dimension to planning. Some people profit from the acquisition, processing, and marketing from resources, and they have an economic reason to continue and sometimes expand resource use. For resources that are considered renewable, planning provides a schedule to use resources at the same pace as they develop. This prevents economic upheaval from cyclical inputs of the resource, and if done properly averages the geographical impacts of the resource.

Water

Water is a natural resource that humans must have constant access to. While we divert and purify water for domestic consumption, the essential planning element is storage. Artificially storing water and/or tapping into natural storage are essential for having a consistent supply. Broadly speaking, Texas has two types of water: surface and ground. Surface water includes both stream flow and impouded water in reservoirs. Ground water originates from aquifers. Under Texas law, surface water and ground water are treated differently. Deriving from Spanish law, the state of Texas owns all surface water. Except for minimal household and on-farm livestock use, a permit is required to use and alter surface water. Ground water is similar to mineral rights; a property owner can pump as much water as they can for beneficial purposes, which is known as the "right of capture" principal. In addition to administering its surface water, the state is facilitating the creation of water conservation districts to manage ground water.

Surface Water

Normal stream flow in Texas rivers fluctuates dramatically between seasons and between years because of our climate. The response has been to construct dams along rivers which backs up the flow of water. The stored water increases during wet periods and then draws down during dry periods. It is argued that dams also provide flood control protection as well as recreational opportunities on the stored water. The stored water is properly called a reservoir, yet many of these bodies of water are called "lakes."

Texas does have numerous natural lakes, but most are very small and intermittent, so they are typically dry. Along meandering rivers in East Texas and the lower Rio Grande, there are "horse-shoe" lakes which were former river channels that became cut-off from the main stream flow. In the Valley, these lakes are called "resacas." In West Texas and on the Panhandle, numerous playa lakes exist when sufficient precipitation occurs. *Big Lake* is an example of one of these dry lakes. The problem of identifying the largest natural lake is that people have modified waterways for storage to the point of not being able to distinguish between natural and artificial. *Caddo Lake* is typically mentioned as Texas' largest natural (and sometimes only) lake. Caddo Lake was formed by a massive log jam that was cleared to allow steamboat traffic. After the lake drained out, a check dam was constructed (in Louisiana) that brought the lake up to its current level. Another augmented natural lake, *Lake o' the Pines,* is upstream from Caddo Lake. Sabine Lake is more properly classified as a bay of the Gulf of Mexico; the flow of fresh water from the Sabine and Neches Rivers provides Sabine Lake with some freshwater characteristics. *Green*

TABLE 8-2 Major Reservoirs in Texas

Reservoir	Location (river)	Area (acres)	Volume (acre/ft)
Toledo Bend	Sabine	181,600	4,472,900
Amistad	Rio Grande	64,900	3,151,267
Texoma	Red	74,686	2,516,225
Falcon	Rio Grande	86,843	2,653,636
Sam Rayburn	Angelina	114,500	2,857,076
Livingston	Trinity	83,277	1,741,867
Travis	Colorado	18,622	1,132,172
Richland-Chambers	Richland Creek	41,356	1,103,816
TOTAL		1,678,708	40,947,816

Sources: Texas Water Development Board; Texas Almanac.

TABLE 8-3 Major Aquifers in Texas

Aquifer	Location	Counties
Ogallala	High Plains	46
Gulf Coast	Coastal Margin	54
Edwards	Balcones Fault Zone	9
Trinity	Metroplex	55
Edwards-Trinity	Edwards Plateau	38
Carrizo-Wilcox	Interior Coastal Plain	60
Seymour	Osage Plain	22
Hueco Bolson	Upper Valley	2
Pecos Alluvium	Toyah Basin	6

Sources: Texas Water Development Board; Texas Almanac.

Lake in Calhoun County, which is adjacent to the Gulf while remaining fresh, is now considered by some as the largest natural lake.

Reservoirs

The debate may never end, but the real king of surface water storage is artificial lakes or reservoirs. Almost 2 percent of the surface of Texas is covered by "inland water" that includes not only reservoirs but lakes, ponds, tanks, etc. With just under 5,000 square miles of inland water, Texas ranks first out of the lower 48 states ahead of Minnesota. In the last hundred years, Texas went from very few reservoirs with little storage to over 6,000 reservoirs. Now we have 204 major reservoirs—defined as a reservoir with over 5,000 acre/feet of storage. Seven modern reservoirs have capacities with over a million acre/feet. In total, Texas has nearly 43 million acre/feet of water storage capacity. The geographic pattern of surface water storage approximates the climatic pattern with much of the total reservation capacity being located in the humid eastern side of the state. All reservoirs are associated with a specific stream channel, and the major reservoirs are located along major rivers.

Ground Water

Deep beneath the surface, another form of water storage occurs naturally in aquifers. Humans can tap into this source of water, but they cannot do much to promote recharging of the aquifer. Texas currently acquires around 60 percent of its water needs from aquifers. The planning around aquifer use concerns sustained usage rates and environmental protection of the recharge zones. Aquifers hold a finite amount of water—both in terms of actual and theoretically possible. The amount of water that can enter or recharge the aquifer is defined by the exposed area of the aquifer and the amounts of precipitation that occurs there. Then, water moves downward with gravitational forces creating a saturated zone, and if enough pressure develops water can pass horizontally or vertically through springs and wells without pumping. The planning imperative is to prevent continual overdrawing of an aquifer, which is the unsustainable use of water that exceeds the natural recharge. The second major concern is the quality of the water entering the aquifer. After identifying the outcrop or recharge area of the aquifer, different land uses and environmental regulations can be instituted, so polluted water doesn't enter the aquifer.

About 81 percent of the surface area of Texas has an aquifer underneath it. Nearly all the major cities and large places have aquifer water to draw on. Although, the biggest user of aquifer water is agriculture; put another way, modern agriculture relies on aquifers for almost 80 percent of what it

178 Part Three Physical Geographies of Texas

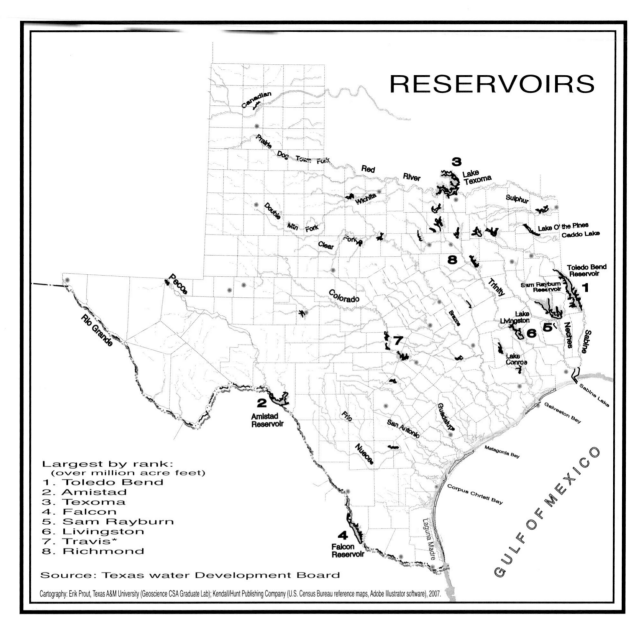

FIGURE 8-6 Texas Reservoirs

uses. The geographic pattern to aquifers is different than surface water. Although river basins appear to be linear features draining toward the Gulf, aquifers more approximate geological patterns.

The Ogallala Aquifer receives a lot of discussion today because it is clear that the aquifer is being overused. One-sixth of all Texas counties sit over the Ogallala formation. Over 90 percent of the water is used for agriculture—mainly irrigating crops. With so much drawdown, concentrations of fluoride and arsenic are increasing. Along the coastline different problems exist with the Gulf Coast Aquifer. Pumping water out of the aquifer creates spaces that can accelerate compaction of the sediments. Close to the Gulf, the removal of freshwater changes the pressure gradients which allows for seawater intrusion. Aquifers are an essential component to Texas' water consumption, and with wise use, they will continue to do so.

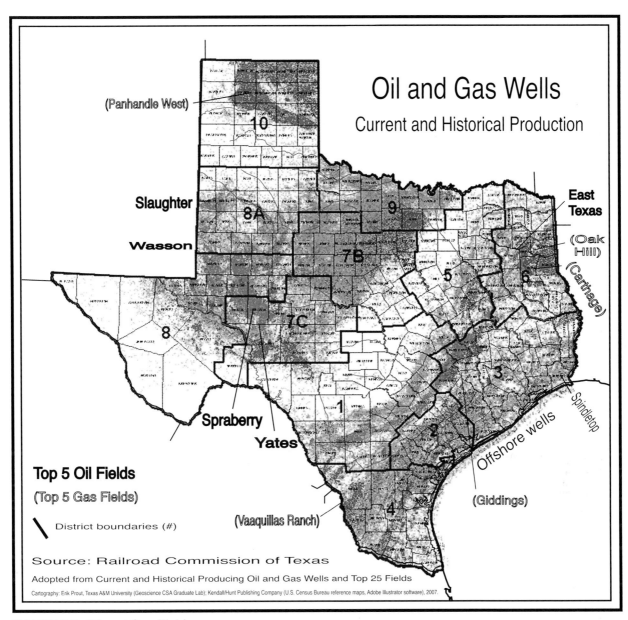

FIGURE 8-7 Oil and Gas Fields

Oil

Oil has had a tremendous impact on Texas in so many different ways. Foremost, oil is a valuable resource that generally has a high value despite fluctuations in market prices. Oil is at its core an energy source, but there are numerous byproducts from petroleum such as lubrications and plastics. And for all practical points, oil has been at the center of the world economy for the past hundred years. While the modern economy would have come to Texas regardless of oil being located here, oil has propelled associated industrial and financial developments. Another consideration of oil is that the economic cycles of oil may not be in sync with the larger American economy. During the 1970s as the oil embargo was creating

FIGURE 8-8 Oil field equipment, Rising Star
Pumping equipment retail facility, Eastland County. Photo by Author.

recession, Texas oil was extremely valuable and driving the state's economy upward. Over time, Texas economy has diversified and oil has become a smaller segment, so most Texans now curse higher gas prices like their fellow Americans.

Spindletop is celebrated as the landmark discovery in 1901 despite earlier discoveries in Nacogdoches and Corsicana. The gusher at Spindletop captured world imagination as well as setting off a wave of oil prospectors. In reality, Spindletop drove the price of oil down to new lows because it increased capacity significantly. Resources such as oil are often associated with these boom/bust cycles. The need for oil during World War II propelled significant federal investment into the East Texas fields (Kilgore) and processing capabilities of Houston and the Southeast. After the war, the West Texas/ Permian Basin fields became the center of oil production; meanwhile, the Gulf Coast becomes a major destination for imported oil. Later, interest developed offshore in the shallow waters of the Gulf of Mexico, so the facilities to support offshore drilling pop up along the Texas coast.

Fossil Fuels

Oil is but one component of fossil fuels. The three basic types of fossil fuels are petroleum oil, natural gas, and coal. Texas has all three types, but coal is not economically feasible. Fossil fuels are primarily the result of biological material being transformed by the pressure of being buried into hydrocarbons. Hydrocarbons take a long time to develop naturally—along geological time scales. Human appropriation of this resource is relatively instantaneous along this timescale. So for all practical considerations, fossil fuels are a finite resource because their creation and depletion are not coherent with each other. Therefore, we treat fossil fuel consumption as an unsustainable form of resource use. To utilize the potential energy, humans primarily burn fossil fuels, which releases carbon into the atmosphere. Without a doubt, burning fossil fuels has been a major component of global warming.

Oil is a good example of how a resource supports a variety of associated activities. The experience of exploring for oil deposits and operating oil drilling equipment has a corresponding human knowledge that transfers to other places. Texas oil companies were able to export their expertise and expand their businesses. In terms of exploration, the value of oil drove geologic knowledge and shaped academic departments. The tangential activities of a legal and financial nature also become a part of the Texas economy. Specialty skills such as oil rig fire fighting and the emerging deep sea drilling are centered here in Texas. The large petrochemical industry in Texas evolved from simple processing of oil. Oil in Texas produced a wide spectrum of economic activities and indirectly prepared Texas for alternative energy endeavors.

Alternative Energy

Despite their finite nature, energy from fossil fuels has played a crucial role in the global economy. Moving to sustainable energy will be a monumental task because of the scale of use and overall dependency on fossil fuels. The alternative energy sources listed below have potential here in Texas as well as elsewhere. Solar energy converts sunlight into electricity. Texas has very high potential in the western side of the state because of the high number of cloud free days. Wind energy converts atmospheric motion (wind) into electricity. The pattern of wind energy potential is very sensitive to local topography yet a series of highly profitable spots are located in West Texas. Recent news stories have boasted that Texas just passed California as the number one wind energy producer. Texas also has some hydroelectric and geothermal potential. Hydroelectric is the production of electricity from the release of water behind a dam, but most reservoirs in Texas do not have long enough drops for large-scale production. Geothermal electricity is produced from the capture of steam after injected water is exposed to very hot rocks. Energy production in Texas has the potential to increase because alternative energy growth could outpace the declines of fossil fuel. With the large petro-chemical industry, it's hard to imagine Texas without an energy industry.

Land Use

Oil and water capture most of the attention when discussing natural resources, yet the most widespread phenomenon is land use. Land use includes agriculture and other primary activities. As mentioned previously, over half of the state is officially categorized as ranchland. In addition to livestock ranching, traditional crop farming makes up this massive land use called agriculture. Agriculture is a natural resource because farmers and ranchers are managing the environment and society shares the benefits of this activity. While some practices are not environmentally ideal, such as biocides, agriculturalists have long-term interests and intimate knowledge of their property. These agriculturally minded property owners are the frontline of soil and water conservation.

Other primary activities such as forestry, fishing, and mining can be added to this land use category. When compared to agriculture, forestry would be the fifth most valuable commodity in Texas. Forestry is a regional resource in Texas. Only in East Texas is a viable, sustainable forestry possible. Reinforcing the regional status, forestry is the number one product in East Texas. The associated activities of lumbering, paper and pulp production, and even a little wood manufacturing is also possible. While not in the popular imagination, Texas has a fishing culture along the Gulf Coast. While commercial fishing has experienced some decline and much transition, it still is significant for many counties along the coast. The emergence of fish farming is just beginning to take shape and will become more significant in the future. Mining on the other hand appears to have declined. Definitely, the mining for strategic or rare minerals like uranium and silver has ceased, but the mining of everyday building materials has increased. With a growing population, construction materials such as cement and crushed gravel and stone are nearly a 2 billion dollar per year activity, which is more valuable than cotton for example.

Conservation/Recreation

The final category of natural resources is a catch-all basket for relatively newer land uses at least from a planning perspective. In reality many of these activities are very old human activities associated with deep convictions and beliefs. As part of an environmental ethos, parts of the Earth are being designated as special places, and these places will have different relationships with people. The conservation movements crowning achievement was the creation of National Park System; National Parks began with the mandate to protect unique natural places such as the Grand Canyon and Yellowstone. In Texas, Big Bend, Big Thicket, Guadalupe Range, and Padre Island Seashore are all part of the National Park System. The federal government with the invitation of the state purchased land to create four National Forests and five National Grasslands in Texas. These areas are generally managed for multiple use that includes both resource extraction and recreation.

Preservation is a slightly different doctrine that originally meant no human activities so nature could evolve without human intrusion. The 17 National Wildlife Refuges in Texas developed out of this idea, yet many are managed similarly to forests with permissible human activities. The state has also purchased and designated land with environmental conservation in mind. There are five state forests and over 50 wildlife management areas.

One of the features of modern life is leisure time. With the advent of a defined workweek and disposable income, people can choose what to do with their leisure time. Many people want to travel and experience the unique features of the state, and many of those activities are associated with the perceptions of nature. Recreation becomes a natural resource because the maintenance of ecosystems and "natural" experiences has this popular purpose. The state park system is quite extensive with many of the parks having a natural dimension, such as Balmorhea and Palo Duro Canyon as well as those designated natural areas like Enchanted Rock and Lost Maples. When one tabulates all the specially designated areas of the state, hundreds of land set-asides are located in Texas.

NATURAL HAZARDS

Natural hazards include environmental elements that cause mortality. A hazard doesn't have to kill people, it only needs to cause harm or damage to people or property. Many hazards are only potentially dangerous, and normally, they cause little disruption to one's quality of life. However, on rare occasions certain hazards cause immense loss of life and damage. Much of the early hazards research originated in harsh unique environments like the Swiss Alps where mountain villages are exposed to avalanches. In the United States, we tend to think of high profile costly events such as earthquakes in California and hurricanes in the Southeast. While hurricane Katrina is still fresh in our minds, many less spectacular events occur almost daily.

Planning for natural hazards cannot really prevent them from occurring; planning works toward mitigation and recovery. Mitigation refers to minimizing the risk, for example, reducing human presence in a dangerous zone or designing better buildings to withstand a specific hazard. For example, houses in California are built not to survive a strong earthquake, but not to collapse on the occupants during the shaking. It is common in many coastal areas to build houses above reasonable storm surges and have shutters to protect against wind damage. Recovery refers to the ability to respond and bounce back from the hazard. Stockpiling emergency supplies, training first responders, and dispensing satellite maps are all examples of recovery that can be done ahead of time. Currently, some elements of hazards research are being applied to counterterrorism; for example, how can we reduce the damage and recover quickly from a terrorist incident? Considering the potentially long list of (natural) hazards, one should employ some organizational structure. For purposes of discussion, natural hazards in Texas are organized into three categories: geologic, climatic, and anthropogenic.

Geologic

Geological hazards originate in the lithosphere—typically at or near the Earth's surface. The spectacular hazardous events are often associated with tectonic plate movements and mountain-building activities because of the sheer amounts of energy and mass involved. With modern photography, most

FIGURE 8-9 Subsidence in Brownwood Subdivision
Ruins of suburban development abandoned due to severe subsidence, Harris County. Photo by Author.

people have seen images of volcanoes. Currently, there are no active volcanic sites in Texas. We often associate *earthquakes* with the West Coast along the Pacific Ring of Fire, but they are very common in Texas. Most earthquakes in Texas are low intensity, translating into not felt, and they correlate well with geological fault lines of the Intermontane and along the Balcones Fault Zone. There is some speculation that a major earthquake fault exists in the Mississippi River valley, so we should not rule anything out.

The majority of geological hazards in Texas are gravity driven, and the presence of water often facilitates the movements. As a tenet of Newtonian physics, the force of gravity acts on everything according to a formula. Whenever Earth materials loose their resistance or friction drops below a certain level, the material will move downward. *Landslides* are one dramatic example of this mass wasting event caused by gravity. Landslides do require some topographic relief, so landslides in Texas are primarily in the far western part of the state. At a smaller scale, the Balcones canyonlands and the Central Texas Uplift also have small landslides and earth movements.

Surface subsidence and sinking are more common around the state. The deep layers and wedges of sediment especially along the Gulf Coastal Plain are prone to compaction. As the weight of new sediment layers on top of older sediments increases, the deeper sediments compact and can eventually become sedimentary rock. *Subsidence* can be exasperated by the pumping of oil and water from those same sedimentary layers where compaction is occurring. Areas around Houston have experienced surface level drops associated with subsidence in the range of 5–6 feet. *Sinkholes* are a smaller scale form of subsidence but the sinking or dropping is more dramatic in terms of depth. Most sinkholes are associated with limestone surfaces such as the Edwards Plateau because of the chemical erosion that occurs subsurface. The sinkholes of the Edwards Plateau correspond with caverns and canyons topography and are important to the recharge of the aquifer. The most common nuisance, if really a hazard, is *shrink-swell* soils with high clay content. Millions of homes in Texas that were built on top of shrink-swell soils have cracked foundations and driveways.

184 Part Three Physical Geographies of Texas

FIGURE 8-10 Flooding Guadalupe and Nueces Rivers, 2003
Texas State Highway 35, Refugio-Calhoun County line. Bottom: FM 666 in San Patricio, San Patricio County. Photo by Author.

Climatic

Climatic hazards derive from the atmosphere. Most of the hazards are associated with storms and/or forms of precipitation. Quite an array of potentially dangerous elements such as wind, hail, and lightning can be present during a single thunderstorm for example. Tornadoes are weather related hazards that are more prevalent in the northern half of the state. Typically tornadoes form along strong air mass boundaries where air begins to rotate as they are mixing; at the surface a tornado is an intense low pressure system with a counter-clockwise rotation. The sheer speed of rotation, strong pressure gradient, and debris that it picks up becomes a potent force that most buildings cannot withstand. Human responses to tornadoes include early warning sirens so people can take cover and storm shelters or basements that are below ground level.

Two distinct climatic hazards are related to each other in terms of the opposite extremes of precipitation. A humorous saying goes as such: there are two types of rain in Texas; too damn much and too damn little. *Floods* and *droughts* are both very costly climatic hazards in Texas and elsewhere. Worldwide, floods kill thousands of people every year and are easily the most deadly hazard. This high mortality is indicative of the human–environment relationship because of the proximity of settlement and agriculture with flood prone areas. One type of flooding is fluvial, which refers to river flooding. During abnormally high precipitation, the amount of water flowing exceeds the channel capacity, and then water and sediment spill over the banks and spread out over flood plain. A common experience in Texas is large areas of standing or moving water (overland flow) that covers roadways during and after strong rains. The second type of flooding is coastal. During a storm event with increased sea level, the coastline literally moves inland and inundates the adjacent areas with seawater and wave energy. Barrier islands and mainland areas less than ten feet above mean sea level are very susceptible to coastal flooding.

Droughts are actually more common and widespread than floods in Texas. While they rarely cause human mortality on their own, droughts are expensive, and by many accounts—the most expensive natural hazard in Texas. The definition of a drought differs between colloquial and professional users. Technically, a drought is a multi-year sequence of dry years with a dry year defined as a percentage below some statistical norm. In Texas, it is common to have dry months, years, and even decades, yet many people project drought onto any dry period. Regardless of the definition, droughts are a hazard because of the high cost of crop failure, the toll it takes on animals, and the resulting change of human interactions with the land.

The big hazardous event for Texas and the Gulf Coast is extra-tropical storms. These are cyclonic storms that form over the tropical oceans and move toward the mid-latitudes. In the North Atlantic, we call these storms hurricanes; they are called typhoons in the Pacific and just tropical cyclones in the Indian Ocean. As they form, they become strong low-pressure systems that develop a counter-clockwise rotation in the Northern Hemisphere. In the Atlantic, the storms generally track to the northwest toward the Caribbean and Gulf of Mexico before turning northeast. Only the storms that enter the Gulf will dramatically affect Texas. Hurricanes produce multiple hazardous elements: bands of thunderstorms with high winds, high precipitation, and lightning; both fluvial and coastal flooding, and storm surges along the coastline. Arguably the most deadly natural hazard in the United States was the 1900 Galveston hurricane, and the costliest was hurricane Katrina in 2005, which is still being tabulated.

Climatic hazards also include extreme temperature events. Temperatures that fall below freezing as well as frozen precipitation cause numerous nuisance hazards such as frozen pipes, dangerous driving conditions, and ice accumulations. Most of Texas experiences a hard freeze and/or snowfall, but there is a strong north-south pattern to frequency. Extreme hot weather events also occur nearly everywhere in the state. Often prolonged heat coincides with drought in the summer half of the year. Heat waves are known to temporarily increase mortality rates for elderly and ill components of the population. Finally, fire is a difficult to categorize as a hazard. When lightning starts a fire, it is a natural (climatic) phenomenon, but

FIGURE 8-11 Hurricane Ike damage
Building across from county courthouse in Anahuac, Chambers County. Photo by Author.

humans also start many fires. The dynamics of how a fire will burn is closely related to climatic conditions such as wind, current rates of humidity, and the state of vegetation. Fire creates an immediate hazard for life and property as well as wildlife. Like with most other natural hazards, ecosystems respond and evolve toward a balance or steady-state that incorporates fire.

Anthropogenic

There is a class of hazards that we call environmental, yet they are not truly natural. A wildfire started by an arsonist that burns the forest around your neighborhood affects what you see, smell, and breathe for months. In this instance, environment goes back to its original meaning of surroundings. A degradation of your surroundings (also a change to "nature") may be hazardous: for example, breathing in ash from the fire may contribute to a future lung disease. Anthropogenic hazards are those modifications to the environment that eventually impact people directly and indirectly, the coherence or balance of the ecosystem people inhabit.

Pollution is the primary label given to anthropogenic environmental changes. Pollution is technically any discarded byproduct of processing and all the discarded material after its useful purpose. Most of these items are benign, and they harmlessly accumulate in landfills. However, modern production and improper disposal has the ability to create extremely concentrated amounts of toxic and dangerous substances. Most people only think of the dangerous materials as pollution, and rightfully, they should dominate discussion about environmental degradation. Pollution can occur in the air, water, and soil, and each has different exposure factors for human health. The geography of hazardous pollution tends to correspond with the sites of production and storage; most of these sites have been identified and are being addressed.

Global warming is a hot topic in the media who plays on the severity of possible consequences. Human modification of the atmosphere from the burning of fossil fuels has already occurred. The increased concentrations of carbon dioxide, methane, and other greenhouse gases have resulted in higher average temperatures measured globally. Global warming is a hazard because the altering of climatic patterns and changing sea levels creates a different and unknown human–environment relationship. Over the short term, some areas will be warmer and/or receive more precipitation and others will not, which will alter food production regions. Some argue that variability and extreme events will increase; therefore, agricultural response may not be able to keep up with changing environments. Security in the areas of food, disease, and disaster response will all be different; for example, health security diminishes as tropical diseases diffuse to parts of the mid-latitudes. We can hope for the best, but we should be planning for the worst.

ENVIRONMENTALISM

Before concluding this chapter on the human–environment relationship, it is relevant to discuss the political and philosophical dimensions to environmentalism. Rightfully or wrongfully, Texas has been scrutinized because of national politics; three recent presidents have officially resided here. Especially during campaigns, their environmental positions and records were examined in relationship to the state. Political rhetoric tends to simplify complex issues, and some of the charges are based on perceptions or worse stereotypes of Texas. On the other hand, many Texans are quick to defend their fellow Texan, which translates into turning a blind eye toward real concerns.

The record of Texas' environment is mixed. There are serious elements of concern as well as things that are going well and worthy of boasting. Foremost, most of our environmental problems are interrelated with being American and living in a developed region of the world. We consume massive amounts of resources compared to the underdeveloped regions. Our advanced economy encourages conspicuous consumption; we purchase symbols to communicate our identity. As Americans, we have an automobile culture that celebrates individual mobility but requires excessive amounts of energy. Moreover, we continue to design our cities and suburbs with automobiles as a priority that perpetuates high energy use into the immediate future.

America as a whole has also been responsive to the idea of studying the environment as well as instituting innovative strategies. Universities in the United States easily spend more money researching the environment than the rest of the world combined. The federal government enacted laws in the early decades of the twentieth century that created the national park system and national forests. Most European countries created only quite recently national parks and ecosystem/biosphere reserves, and the size of most specially designated areas are quite small when compared to the United States. In the 1970s, Congress passed the Endangered Species Act, which was the groundbreaking legislation that much of contemporary environmental protection is based on. Since then, we have successfully preserved our national symbol—the bald eagle, which is the highest profile success. However, many other aspects such as the spotted owl/old growth forest and the reintroduction of gray wolves in Yellowstone have been controversial.

Perhaps, the big question for Texans is how we see our role toward the environment. Do we want to be active participants for conservation or do we want to be passive observers of degradation? Gunter and Oelschlaeger present a similar question in their book, *Texas Land Ethics*. The term *land ethic* comes from Aldo Leopold who inspired many environmentalists with his writings. Leopold argued that all ethics refer to the balance between the individual and the community; the individual competes for resources but also cooperates because she needs the community to exist as an individual. Land ethics expands the realm of community to the land (and plants, animals, rivers, etc.) from which the individual derives natural resources. The informed citizen with land ethics seeks to prevent degradation, repair previous damage, and mitigate future transgressions. It is a tall order that requires the whole community. It also requires an understanding of the scientific research on the physical world. In this part, we discussed the physical geography of Texas that illuminated some of that science.

"Texas Is an Urban Place"

FIGURE 9-1 Houston CBD Skyline
The Central Business District/ Downtown from the Montrose District, Harris County. Photo by Author.

Part Four
Modern Human Geographies of Texas

INTRODUCTION TO CONTEMPORARY HUMAN GEOGRAPHIES

The contemporary human geography of modern Texas revolves around its growing population. Almost every topic of discussion such as the economy and environment is affected by population growth. As Texas' population grows, its development level also rises. The introduction of technology combined with increasing energy consumption has created new geographies of human interaction and presence on the Earth. This process of change doesn't dramatically begin on any specific date; the adaptation of technology that transformed agriculture and initiated manufacturing was more continuous in nature. Yet, the oil gusher at Spindletop is often presented as a defining moment because the oil based economy that comes afterward is truly a 20th-century phenomenon. One dimension of the oil economy that utterly transformed Texas is the automobile. It's unimaginable to discuss modern Texas without considering the impact of transportation. First came the railroad that economically integrated the country, and then the highway was constructed to accommodate the automobile.

With all this change to the state, the astute question to ask is what new geographies were created. Does location still matter? Moreover, can we observe structure and organization to these changes? It may not be as simple as observing place-names and eco-tones in the landscape, but yes modern Texas does have patterns that reflect the forces of democracy, capitalism, demography, and civil society. Governmental institutions are an integral part of modern Texas because they indirectly mediate the average person's experience with large scale changes for the collective good, hopefully. For example, governing 25 million people and managing 268,000 square miles of lands takes both structure and organization; numerous administrative districts are necessary to perform properly.

This part on modern Texas has only three chapters because we need to generalize as we survey so many topics. First, I introduce the idea of modern as it pertains to the contemporary cultural landscapes. Chapter Nine examines the demographic changes in terms of population growth and the dynamic cities of Texas that are absorbing the growth. Chapter Ten discusses some of the societal structures that underpin our modern lives. We explore the political and economic geographies of Texas. Finally, Chapter Eleven explores the role of leisure—including recreation—into shaping the contemporary cultural landscapes.

MODERNITY

For many people, "modern" means the present: right now, and not in the past. So it must seem strange when academics and others such as artists, architects, and planners refer to something as "post-modern." How can something be post-present? Isn't that the future? For those educated to the idea, modernity is the ultimate manifestation of the Enlightenment. The idea that rationale thought and in particular science would elevate humanity to a higher level. For those designing and constructing buildings, highways, landscapes, and even public art during those decades, they were both contemporary and modern. Perhaps, the ultimate symbol of modernity is the skyscraper that dominates urban landscapes around the world.

The "modern" skyscraper is physically possible because of the internal steel frame; previously buildings had to have load-bearing walls. In addition, little technological feats such as elevators, air circulation, and water supply had to be worked out. As a part of this modernization ideal, the buildings themselves were very simple and efficient in appearance; most were square in both a vertical sense as well as a literal footprint on the surface. The square footprint was meant to maximize the use of the expensive property they are built on, so there is a rational financial dimension. The square shapes are attributed by the pre-existing grid pattern of city streets. The rigid rectangular was both a visual aesthetic—usually labeled modern—and a rational way to maximize the economic value in terms of square footage of rentable space above ground level. Even though many modern buildings have standardized floor sizes, the actual floor plans were extremely flexible and could be configured into many different patterns of offices and cubicles for example.

Modern buildings such as skyscrapers tended to be similar to one another, and they really had no regional or place specific qualities. Residential houses, shopping malls, and commercial strips also tended to be placeless because they were increasingly standardized with common designs and similar construction materials and techniques. The prefabrication of homes and box stores exemplifies this to the extreme. Fast food chains provide another sort of example to this modern landscape discussion. On one hand, they want to be extremely different with a visually distinct architecture and landscaping that consumers will recognize. On the other hand, they replicate this distinct look everywhere they locate a new establishment.

Post-modern includes the criticism of the rational use of space and standardization of landscapes that seem to have become so dominant in the cultural landscape. At the intellectual level, post-modernity is a rejection of the rigid constraints of rational thought. There are multiple paths and not one single truth. Marxists identify post-modernism as the later stages of global capitalism. At the other end of the spectrum, post-modernity is about personalization of space and recognition of place. Santa Fe and New Orleans can be unique with their own architecture and landscape preferences. While some of the aesthetics appear to be pastiche, they genuinely try to identify with the local senses of place.

Part Four Modern Human Geographies of Texas   191

LEARNING OBJECTIVES: MODERN HUMAN GEOGRAPHY

Part Four of this textbook is associated with modern Texas and the institutions and pratices of contemporary society.

- Discuss the contemporary human geography of Texas.
  You should be able to discuss where people live, work, and play.
  You should be able to relate cities with interactions, goods, and services.

- Identify and elaborate on the structure and organization of modern Texas.
  You should be able to provide examples.

- Know the locations of the major metropolitan areas and highways of Texas.

- *Maps:* Familiarity with modern maps of Texas.
  Identify and interpret demographic and economic maps.
  Read and use classroom examples: political regions, economic regions, and tourist regions; interstate highways; Metropolitan Statistical Areas of Texas.

- *Definitions:*
  Modern, Modernity, Modernization, and Post-Modern
  Demography
  Urbanization
  Partisan and Ideological Politics
  Development (uneven)
  Infrastructure
  Leisure and Recreation

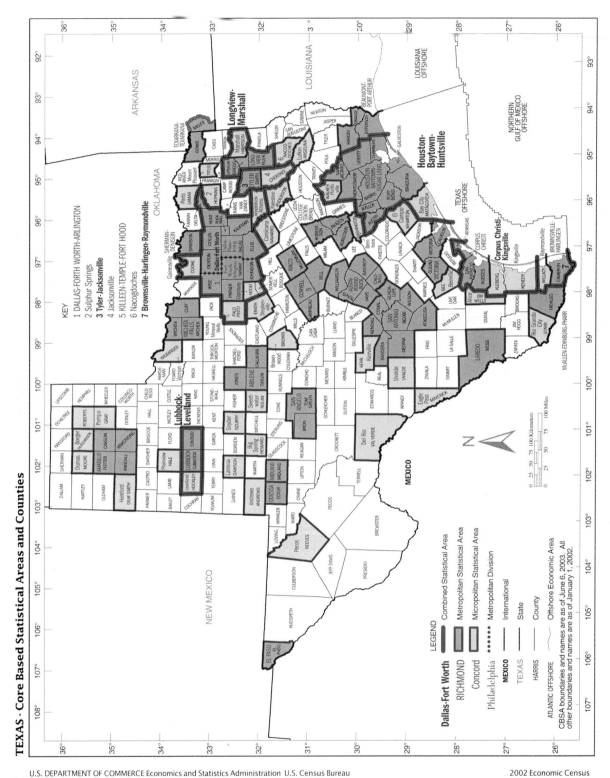

FIGURE 9-2 Metropolitan Statistical Areas
Map courtesy of US Census Bureau.

9
Millions of Texans

Texas is big—not just area but population! As mentioned earlier, Texas is the second most populous state after California. Like other fast growing states, Texas has a lot of cultural diversity. Much of the growth and diversity is increasingly located in a few large cities. Despite the stereotype of wide open spaces and the reality of an agrarian past, Texas is very much an urban place. As we begin to think about modern Texas, two aspects to describe are where all these people are located and who are they? In this chapter, we discuss the population geography and urban geography of Texas as we answer these questions.

Migration has been a consistent element of Texas history. Inevitably, as new people arrived, the cultural balance between the different groups changed. Initially, most migrants were agriculturalists, and much de facto segregation between groups existed in rural Texas because they created their own towns. Over time, rural folk migrated to larger towns and cities for economic and educational opportunities that were culturally more diverse. "Melting pot" may not be the best metaphor, but some mixing and blending occurred. Like the United States as a whole, recent immigrants are arriving from Asia and Latin America. Currently, both foreign and domestic migrants tend to locate directly into urban and suburban areas. In general, Texas cities are extremely diverse on par with other large American cities.

DEMOGRAPHY

Demography is literally people-writing or the study of the population. Various academic disciplines primarily in the social sciences have specialists in demography. Population geographers' niche is the study of the spatial and environmental aspects of human populations. Geography's first contribution is in the form of mapping populations that helps many visualize the spatial aspects of population change. The second contribution is to conceptualize population change in terms of a broader human-environment relationship that highlights ecological degradation. For our purposes of the regional geography of Texas, I elaborate on only a few aspects; they are absolute population, population growth, population characteristics, and population composition.

Before discussing each demographic dimension, let me say a word of caution and context. Demographic data are collected and disseminated by governments. The U.S. Constitution mandates a decennial census for the purpose of congressional representation, so the American people are accustomed to this intrusion. Many people in the world (as well as some Americans) do not have a trustful relationship with their government, and it is quite common for people to misinform census takers. Governments also have interests in how they report census data to the rest of the world. At times governments boost their numbers to imply a larger military potential or to receive some international assistance. Other times they may try to hide embarrassing or inconvenient facts such as AIDS cases and ethnic cleavages. Despite these problems, demographic data from censuses are extremely valuable for understanding human groups.

Population

The first statistic of population is an exact number of how many people are in a defined region such as a state or city. Unfortunately, it is not as easy as a simple head count. Even the U.S. Census has problems with undercounting especially in urban areas. During the last official Census in 2000, the population of Texas was enumerated as 20,851,820. Since then the population has increased to nearly 23 million based on the 2005 estimate. Texas accounts for approximately seven and a half percent of the U.S. population total that very recently surpassed 300 million. At the global scale, the United States accounts for less than 5 percent of the world's 6.5 billion people. Therefore, only about one-third of one percent of the world's population resides in Texas.

Population Change

Immediately after learning the number of people, a second dimension is what sort of growth does that number represent? The proper term is called population change because the number may represent growth or decline. In Texas' case, population change of each decade since statehood has been growth. To be fair, the United States as a whole and most all states exhibit growth every decade. Naturally, some states grow faster than the national average; however, a few states like Texas grow much faster than most. Texas has impressively grown faster than the national average each and every decade since 1850.

There are multiple ways to measure change and growth. The most obvious way is to use the actual numbers to calculate an absolute number. For example, the difference between the 1990 and 2000 censuses is 3,865,310, which means there were nearly 4 million additional people in the state during the 1990s. The absolute number can be stated as a percentage growth rate. For example, Texas' population grew by 22.8 percent during the 1990s decade, which is calculated with 3,865,310 divided by 16,985,510. Demographers prefer statistical measures that can be easily compared to other places and related to other phenomena. The most common is the rate of natural increase (RNI); the RNI is the difference between birth rates and death rates. The RNI is useful because it is directly related to doubling rates and indirectly related to fertility rates.

Population growth in Texas has two components. The first component is the RNI; birth rates exceed death rates in Texas. Moreover, this natural increase tends to be slightly higher than the national average. The second component is migration because more people are moving into Texas than those moving out. Ultimately, total population change of any place is a function of natural increase or decrease and plus or minus net migration. In Texas' case, we typically add both natural increase and positive migration together. Interestingly, the percent of Texas' growth changes between decades as the relative weight of migration varies. In recent decades, migration has been responsible for roughly half of the growth.

TABLE 9-1 Population and Scale, 2010

Geographic Scale	Population (mid-2010)	Percentage of Previous Scale	Pop. Density (square miles)
World	6,868,528,206		117 (land)
United States	308,745,538	4.50% of world	87.4
Texas	25,145,561	8.14% of U.S.	96.3
Brazos County	194,851	0.78% of Texas	265.5

Sources: U.S. Census Bureau; Texas Almanac; Population Reference Bureau.

TABLE 9-2 Population Growth in Texas (Decennial census)

Year	Population	Natural Increase	Migration	Total Growth (absolute)	Percent Growth
1850	212,592				
1860	604,215			391,623	184%
1870	818,579			214,364	35%
1880	1,591,749			773,170	94%
1890	2,235,527			643,778	40%
1900	3,048,710			813,183	36%
1910	3,896,542			847,832	28%
1920	4,663,228			766,686	20%
1930	5,824,715			1,161,487	25%
1940	6,414,824			590,109	10%
1950	7,711,194			1,296,370	20%
1960	9,579,677	1,754,652	113,831	1,868,483	24%
1970	11,196,730	1,402,683	214,370	1,617,053	17%
1980	14,229,191	1,260,794	1,771,667	3,032,461	27%
1990	16,986,510	1,815,670	941,649	2,757,319	19%
2000	20,851,820	1,919,281	1,946,029	3,865,310	23%
2010	25,145,561	—*	—*	4,293,741	21%

Source: U.S. Census Bureau.
*breakdown of growth not available yet.

Population Density

Population density is a measure of how many people live in a defined amount of space. Population density is important because it often correlates with environmental modification and settlement impact. The more people in a small area like in a city have a large footprint by their sheer presence than the opposite scenario of few people in a large area. Furthermore, there must be an organized system of acquiring and delivering resources to large concentrations of people in cities. Population densities are often associated with rates of urbanization in the developed world, yet historic civilizations such as China and Egypt have high densities in agricultural contexts.

Typically, population density is calculated after the census count and converted to known units of space such as square miles or kilometers squared. For Texas, we would divide the current (2005) population estimate of 22.9 million by 261,797 to get 87 people per square mile. With a population density greater than 84, Texas officially has a higher population density than the American average. That's not to say there aren't wide open spaces in Texas. Population density varies widely between places. Large areas of Texas have population densities of less than 10 people per square mile. On the other hand, a few urban counties have over 5,000 people per square mile.

Population Distribution

The distribution of population in Texas has changed over the decades. Distribution is literally the geography of where people are located during the census count. Individuals are assigned to their place of residency. Prior to the Republic, much of the population was located in Spanish missions and towns, with

196 Part Four Modern Human Geographies of Texas

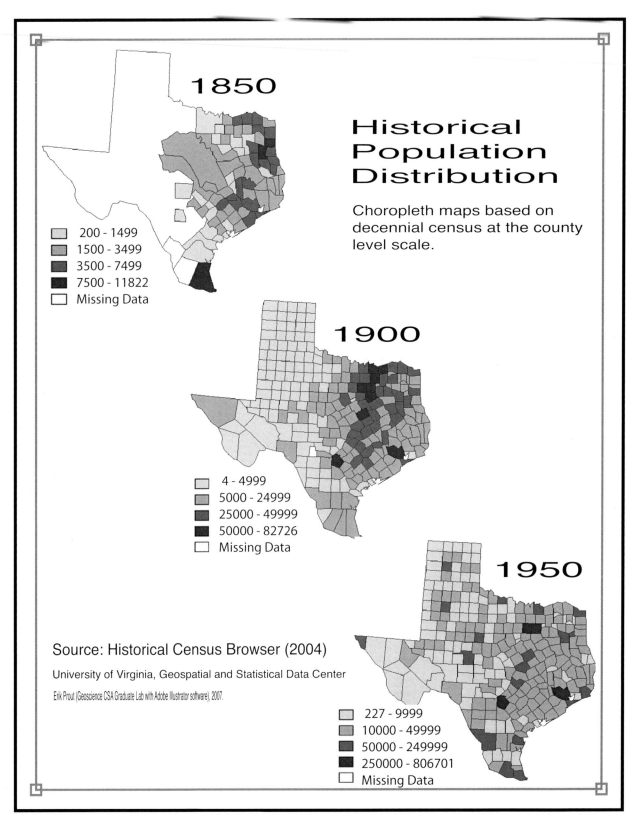

FIGURE 9-3 Population Distribution

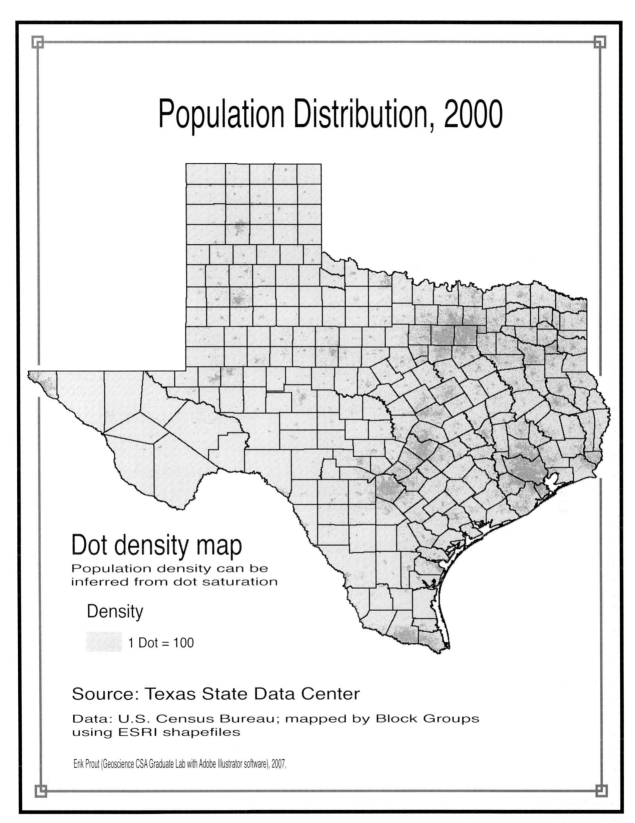

FIGURE 9-3 Continued

San Antonio being the largest place. As the Empresario lands began to be settled, the center of population moved eastward and toward the coast. By the time of the first U.S. Census in 1850, the center of Texas population was in Madison County; basically, half of the population was East of the Trinity River. The center of population moved westward and by the turn of the century Waco was the approximate center of Texas population. During the second half of the 20th century, the center moved slightly southward and currently is located in Bell County.

The more dramatic change in population distribution is the movement of people from rural areas to the fast growing cities. Using the 254 counties of Texas as our scale of analysis, we see some significant types of change. Foremost, Texas population has become extremely concentrated in a few counties. Over half of all Texans reside in just seven counties; Harris and Dallas counties together account for over one-fourth of the population. On the other end of the spectrum, two hundred least populated counties barely account for 15 percent of the population. Over half of all counties in Texas have a smaller population today than they did in the past. While a dozen or so places are experiencing rapid population growth, hundreds of other places are experiencing stagnation and even population decline.

Population Characteristics

Populations are different from one another across all geographical scales. Each population has basic characteristics that can be compared with all the others. Only two generalized characteristics are presented here: structure and composition. Structure includes age and gender, which are common to all populations, but each society creates different meanings from the data. Meanwhile, composition refers to the racial and ethnic categories that each society defines for itself. Immigrant societies like the United States tend to be interested in ancestry and heritage based questions as well as an accounting of citizenship status.

Age/Gender Structure

Every population has an age/gender structure, and this structure is often represented as a population pyramid. Left and right sides of the pyramid are gender—typically with male and female respectively. Age is shown vertically with young at the bottom and old at the top; often the top bar is a greater than category. If the number of live births was constant, the shape of a population pyramid would be relatively smooth gradually becoming wider at the bottom. That doesn't happen. Birth rates fluctuate and those years with a baby boom are visible as they age. For a pyramid based on the 2000 Census, the 35- to 44-year-old age groups (Baby Boomers) are wider than age groups (Generation X) beneath them.

Both age and gender are discussed separately as averages. Median age is the age that half of the population is older and half is younger. The median age for the United States was 36.4 years in 2005. Texas is slightly younger at 33.2, but that represents the second youngest population after Utah. As the population ages in America, the median age climbs slightly upward. The best explanation for Texas' lower median age is the influx of migrants. Immigrants, especially short-term employment seekers, tend to be young adults. On a lesser scale, the influx of 20- and 30-year-olds are also in their prime child-bearing years, so Texas has a slightly higher percent of school age children than the American average.

The gender ratio is 97.6:100 in Texas. The ratio is translated as for every 100 females there are 97.6 males. Like the United States average of 96:100, there are more females than males. Texas is slightly more balanced but still more females. Interestingly, more baby boys are born than baby girls, but by their 30s the two are equal. Thereafter, females outnumber males and it becomes very pronounced above age 70 with women's longer life expectancy. Although frontier places such as Alaska (103:100) tend to have more males than females, that is not the likely explanation for Texas's gender ratio ranking. Once again being a migration destination is more significant as the male/female ratio of migrants favors men.

Population in Texas

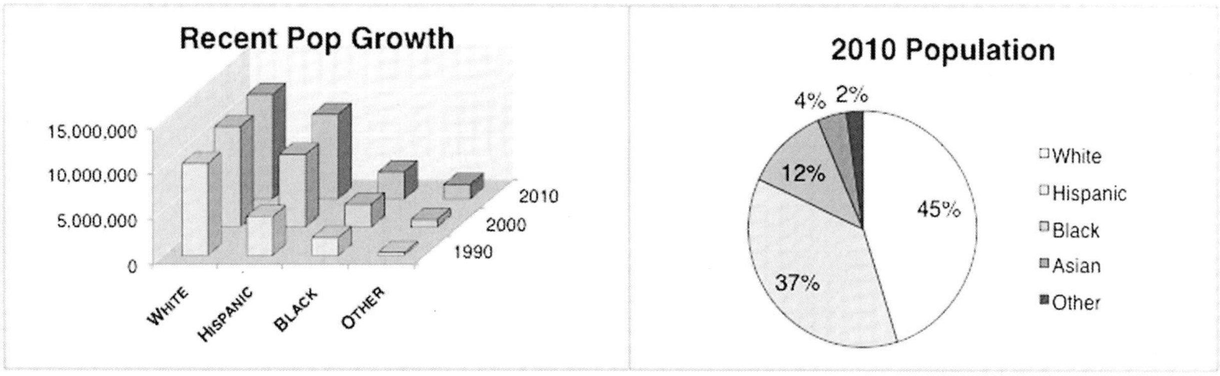

FIGURE 9-4 Population Characteristics
Source: US Census Bureau.

Racial/Ethnic Composition

Further differentiation of the population is more sensitive than age and gender because the categories are socially constructed. In a literal sense, the census takers have to formulate the questions, and the questions reflect each society's prerogatives. For example, does the Iraqi government want to know the ratio between Shia, Sunni, and Kurd? Would knowing the exact ratio help or hinder political solutions? In Lebanon, the constitution mandates a Christian president and a Muslim prime minister. Because no official census has taken place since the 1970s, nobody knows definitively if the religious balance still exists. Even in Europe, Belgium doesn't officially ask language related questions so it can pretend the Flemish (Dutch) to Walloon (French) ratio is equal.

In the United States, we have multiple identity questions that are interwoven, but two stand out. Both the concept of race and ethnic heritage are intellectually disputed ideas, but at the individual scale, they are real phenomena with serious meanings, personal identification, and social consequences. Most scholars completely discredit racial categories because they originated with Eurocentric notions of one race being better than another. Furthermore, modern genetic research tends to prove the contrary position that there is only one human race; we are more alike than different from each other. However, human differentiation does exist, and some of the differences have geographical manifestations. Skin color is one such feature. Groups living in a specific environment for a long enough time will evolve dermatological distinctions. Different skin pigmentation is empirically real, but the jump to categories with connotative meanings is a learned cultural feature.

A layperson would probably use a simplistic, color coded system such as black (African), white (European), brown (Mexican), yellow (Asian), and red (Native Indian). While that would probably work rather well, the biggest problem is with dual categories for Hispanics. The reality in Latin America complicates this task because Spanish colonization successfully created relatively stable mixed race societies. The census-takers dilemma is determining between primary and secondary identities or allowing the respondents to decide. The U.S. Census considers the two distinct enough to ask both. Race is White, African, Asian, and Native American; ethnicity is Hispanic and non-Hispanic. In addition, the Census asks other questions such as place (country) of birth, citizenship status, ancestry association, and predominant language; therefore, a good picture of racial and ethnic diversity in Texas exists.

TABLE 9-3 Population Composition, 2010

Census Category	Population	Percentage of Total Population	% 2000
RACE			
White	14,799,505	70.4	+19.6
Black or African American	3,019,318	12.0	+23.9
American Indian or Alaska Native	178,127	0.7	+44.4
Asian	1,063,715	3.8	+71.5
Native Hawaiian or Pacific Islander	34,506	0.1	+50.0
Some other race	61,466	11.7	+7.8
Two or more races	319,558	1.3	+31.9
ETHNIC			
Hispanic or Latino	9,460,921	37.6	
Not Hispanic or Latino	15,684,640	62.4	+41.8
COMBINATIONS			
Anglo (White not Hispanic)	11,669,272	46.4	+41.8
Other race (not Hispanic)	61,466	0.2	+10.6
Total 2010 Population	**25,145,561**		**+10.6**

Source: U.S. Census Bureau, Census 2000.

The demographic composition of Texas is changing. This change in composition is just as dramatic, if not more so, than the change associated with growth. The two changes are interconnected. Growth of Hispanic outpaces the statewide growth rate, and the growth of Anglo does not keep up with the state average. The relative weight of these two largest groups is reversing. Already Texas has no majority group. Some scholars call this a "Minority-Majority" situation because the collective percentage of non-Anglos (White/non-Hispanic) is over 50 percent. Anglos remain the largest single group, which is known as a plurality. Hispanic population is trending toward becoming the largest group; first it will become the plurality when it passes Anglo and then the majority after exceeding 50 percent.

The consequences of this change are openly discussed. However, some of the discussion is political rhetoric bordering on racism. In addition to Texas, many other regions of the United States are experiencing high rates of immigration from Latin America and elsewhere. Nearly all the counties in the western half of the United States have a Hispanic majority, plurality, or prevalent minority. Counties in the South and Midwest that never had migrants from Mexico and Latin America as a whole now have a fast-growing Hispanic component to their populations. California exemplifies this national trend with over one-fourth of its population being foreign born (including its governor). While also being the dominant destination for Asian migrants, most of the foreign born population in California are still from Mexico. Texas has a very similar demographic profile. Texas has an above average percentage of nearly 16 percent of the population being foreign born, and a vast majority of those were born in Mexico. Migration from Asia is increasing even faster than from Latin America, but the absolute numbers are still smaller. Overall, Texas is one of the most diverse states in the United States. If California wasn't attracting so much of the diversity discussion, American demographers would probably pay more attention to Texas. Texas has the second largest Hispanic and African-American populations in the United States; it also has high rankings of Anglo, Asian, Native American, and many other recent migrant groups. Most of this diversity is best reflected in urban Texas, which we turn to next.

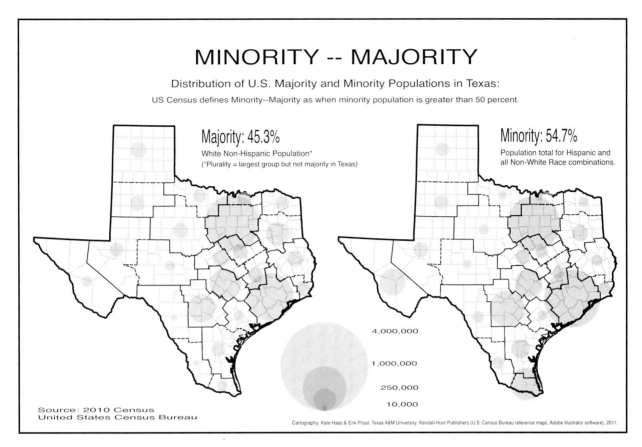

FIGURE 9-5 Minority & Majoriy Texas

CIVITAS

In the course of human history, cities are relatively new environments that came about during the last 10,000 years. As agriculture improved to the point of reliable surplus, complex societies developed in situ. Higher densities of population entailed sophisticated forms of governance such as city-states. The oldest city archaeologists know about is Ur in modern day Iraq. Eventually, great civilizations and empires arose out of certain city-states; examples include Mesopotamia, China, and Mesoamerica. By today's standards, those early cities were very compact typically with defensive walls around the perimeters. A city's centrality extended beyond economic markets and social networks; cities were the centers of power that made them the focus for geopolitical expansion. No one understood this better than the Romans who expanded their empire through the process of "citification." The way to incorporate diverse people into a single civilization was through the conversion of civil culture in cities. In fact, the idea of "civilizing" someone is identical to "citifying" them, so it was only in the city that someone could really become Roman.

Spaniards introduced this Western notion of the city into the Americas during the Columbian Exchange. They founded numerous cities and dictated the initial urban pattern throughout much of their Empire. At the same time they encountered indigenous cities and symbolic places that they transformed into Spanish places. Perhaps, the most monumental transformation was in Mexico City where the Spanish built the main plaza, *Zócalo,* on top of the most sacred Aztec temple. Out on the frontier, Spanish missionaries were creating little communities that replicated the civil order of

Spanish cities. In between the core and periphery, networks of cities were being implanted in the landscape and connected by roads.

Texas Cities

In Texas, the Spanish laid out the framework of roads and towns that shaped early settlement, and then the Mexican government expanded on it during their control. After the Revolution, southerner's ideas about cities became the dominant pattern for development. The Southern colonies had a different city–rural relationship than the Mid-Atlantic and New England colonies. Initially, cities were less important because rural areas tended to be self reliant; in particular, plantations had a cellular structure that traded directly with foreign markets. Small cities formed throughout the interior almost as an extension of the rural crossroads, but a few places, especially gateways, became relatively large. Structurally, cities in the South had to accommodate slavery that required both economic integration and social segregation between blacks and whites. While Charleston, Richmond, and Savanna are early southern urban models, the cities along the Mississippi River including Memphis, Natchez, and New Orleans were actually experienced by Southerners moving to Texas.

All Texas cities began as small places, but only a few arose to great size. The initial pattern was consistent with Spanish design that combined their geopolitical aspirations with the reality of terrain and rivers. At the frontier's edge, *Los Adeas* was founded as close as possible to the French trading posts working their way up the Red River. Despite being a forward capital, Los Adeas and the entire eastern frontier never really attracted Spanish settlers. The long distance from Mexico necessitated way stations and ferry crossings that created intervening opportunities. San Antonio was approximately halfway between Saltillo and Los Adeas, and more importantly it had fresh water springs. After a cluster of missions formed in San Antonio, it was easily the largest place in Texas for over a hundred years. In 1731, a formal city charter was issued to a group of Canary Islanders who settled in San Antonio; they created the first real municipal government in Texas.

Past Rank Orders

With census data, there has always been a largest city. However, Texas never really developed an urban structure where one place dominated over all the other places. No single city became the strong central place that anchored the entire regional economy. Since the 1850 Census, four different cities have claimed the title largest city in Texas. Galveston topped San Antonio on the list in 1850, and only eight other places had over a thousand residents: Houston, Gonzales, Victoria, Marshall, Tyler, Paris, Palestine, and New Braunfels. Only recently did the largest city reach 10 percent of the total state population. The cumulative populations of the second and third largest cities always exceeded the largest city. By the end of the 19th century, a multinodal urban pattern developed. The three dominant urban nodes were San Antonio, Houston/Galveston, and Dallas/Fort Worth.

Historically, San Antonio was the largest place longer than any other city, and it has retained its relatively large population. San Antonio was the first Texas city to reach 100,000 people, and it was officially the largest city in the 1860, 1900, 1910, and 1920 censuses. The mid-19th century growth had much to do with being a gateway to the American Southwest. The Army contributed to San Antonio's growth by using it as a staging point for securing the Mexican border and fighting the Plains Indians. Fortunately, San Antonio was connected to the emerging transcontinental railroad networks. Yet it was the railroad links to Mexico that propelled growth in the early decades of the 20th century. During the middle and later decades of the 20th century, San Antonio trailed Houston and Dallas. San Antonio benefited from military bases, tourism, and an attractive, high quality of life. Recently, San Antonio became the second largest Texas city.

TABLE 9-4 Largest City by Decade

Census Year	Largest City	Population	Next Largest 2/3/4/5
1850	Galveston	4117	SA / H / Pa / NB
1860	San Antonio	8235	G / H / Ma / NB
1870	Galveston	13,818	SA / H / Br / A
1880	Galveston	22,248	SA / H / A / D
1890	Dallas	38,067	SA / G / H / FW
1900	San Antonio	53,321	H / D / G / FW
1910	San Antonio	96,614	D / H / FW / EP
1920	San Antonio	161,379	D / H / FW / EP
1930	Houston	292,352	D / SA / FW / EP
1940	Houston	384,514	D / SA / FW / EP
1950	Houston	596,163	D / SA / FW / A
1960	Houston	938,219	D / SA / FW / EP
1970	Houston	1,232,802	D / SA / FW / EP
1980	Houston	1,594,086	D / SA / EP / FW
1990	Houston	1,637,859	D / SA / EP / A
2000	Houston	1,953,631	D / SA / A / EP
2010	Houston	2,009,451	SA / D / A / FW

Source: Texas Almanac.
SA=San Antonio; H=Houston; G=Galveston; A=Austin; D=Dallas; FW=Fort Worth; EP=El Paso; NB=NewBraunfields; Pa=Palastine; Ma=Marshall.

Galveston Bay became the dominant seaport gateway for Texas after the Revolution. Initially, the city of Galveston capitalized on its natural harbor situation being the best place for ocean going ships to lay anchor. After Indianola was destroyed twice by hurricanes, Galveston was unrivaled as the main Texas port. Numerous European immigrants passed through Galveston—a smaller scale Ellis Island. Galveston was Texas' largest city in 1850, 1870, and 1880. Houston became Galveston's main seaport competition by aggressively building a ship channel and supporting business growth. Houston actually overtook Galveston prior to the 1900 hurricane, and afterward, it was Texas' primary port. Houston's advantage over Galveston was railroad access, which allowed it to be more integrated in the economic changes associated with the railroads. In addition, Houston was able to capitalize on the discovery of oil into a broad economic base. The oil economy boosted its population to become the state's largest city beginning in 1930 and it has remained there to the present. Houston reached numerous urban milestones such as being the first Texas city to reach 1 million and very recently became the only 2 million-resident city.

The remaining node of this regional structure was Dallas. Lacking the obvious locational advantages of being a gateway, Dallas impressively arose to become the largest city in 1890. Thereafter, Dallas was the second-largest city for 10 consecutive census counts. The initial growth of Dallas had much to do with becoming the central marketplace for cotton. Cotton production was transitioning both in terms of location and transportation; Dallas quickly integrated by railroad with the Midwest and Houston. Using the financial experience of cotton markets, Dallas was poised to be a major player in the development of oil in Texas. Like the Houston and Galveston situation, Dallas had regional competition from Fort Worth. Fort Worth tended to look West as it developed a cattle market along with its stockyards, which later evolved into a meat-packing center. Together, Dallas and Fort Worth anchor the region that would become the largest metropolitan area in Texas.

Urbanization

Scholars conceptualize urbanization as more than some population threshold, but more of a social phenomenon. In effect, urbanity or urban culture is a state of mind. Urbanity is the way people in a high-density environment survive. By these measures, Galveston is easily the most urban place in Texas. Because Galveston was the point of immigration for Europeans and even Americans from the East Coast, it had received urbanites from many different parts of the world. These migrants were already familiar with cities and they diffused their urban ideals and notions to Galveston. And from Galveston, urban designs of space such as city parks and alleyways diffused to other Texas towns.

Another scholarly idea is that urbanization in America coincides with industrialization. As the scale of manufacturing and industrial production rise, a concurrent ability to house large numbers of workers is necessary. The rapid growth of cities coincides with these parallel developments. Some scholars argue that industrialization and urbanization are synergetic and they propel each other. For example, skyscrapers require technological advances associated with both the need and ability to produce such an urban form. From a planning perspective, cities must set aside large areas for industrial scale production as well as large areas for residential housing. Furthermore, they must be connected by mass transit or public infrastructure such as highways. The net result is that very large cities tend to have well defined zones with different functions such as manufacturing and residential. Cities also have a way of segregating people along socioeconomic status.

The modern Texas city is very similar to the American model because the same driving force of automobiles is paramount. Yet the early American city began as a small nucleus with everything being within walking distance of one another. The early expansion of cities began with mass transit lines of horse drawn and eventually electric streetcars. These linear lines of development radiated away from the core creating a spoke appearance. When the automobile came about, cities could infill between the spokes and expand along new lines of major boulevards or highways. The interstate highway system (freeways) created new lines of mobility that simultaneously expanded the possible size of the city and also physically divided the city into sections. While the previous highway network invented the bypass, the freeway network has evolved a beltway or loop system that creates a highway ring around the city's periphery. No longer does one have to directly interact with the city center while remaining a part of the urban system.

Metropolitan Statistical Areas

The Census Bureau has created a statistical entity to account and better understand urban America. The *Metropolitan Statistical Area* (MSA) is commonly defined as a freestanding urban entity with over 100,000 people. As urban areas expanded, they often include multiple numbers of cities, and sometimes these areas extend beyond county and state borders. The first geographical consideration is to select the scale of analysis. The MSA is based on county-level data. In much of the country, cities are confined to county boundaries, and the areas in one county are linked to only one metropolitan core. For Texas, this works rather well because the size of counties are similar to the American norm; however, there are instances when cities can be in multiple counties. The second geographical consideration is the determination of freestanding status. When the places in question are separated by rural counties, it is easy to distinguish independence. In and around big urban complexes, there is an implied dependency, yet the Census Bureau uses commuting patterns to confirm the linkages. MSAs follow a labeling hierarchy that is dictated by population size and minimally influenced by other considerations.

Currently, there are 25 MSAs in Texas consisting of 77 counties. Almost all population growth in Texas is in these 77 counties. As mentioned previously, there is a distinct imbalance in county populations.

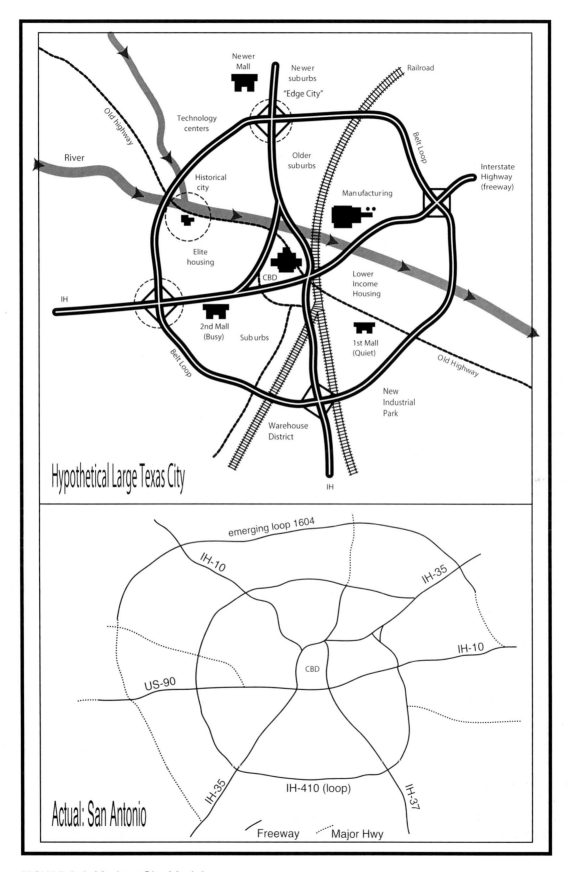

FIGURE 9-6 Modern City Model

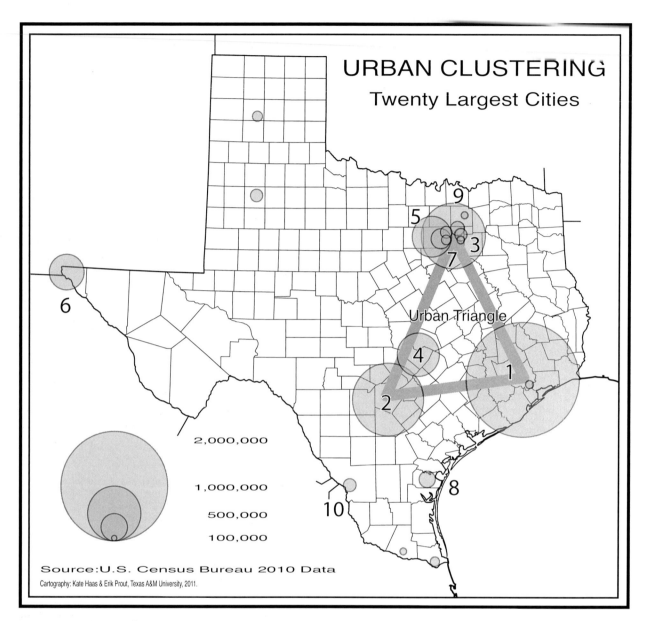

FIGURE 9-7 Urban Clustering

The seven counties that account for over half of the population are Harris, Dallas, Tarrant, Bexar, Travis, El Paso, and Hidalgo. In total, nine counties have over half a million residents, and Harris tops the list with nearly 3.5 million. As we would expect, these populated counties are the core for their respective MSA. Discussion of MSAs highlights one of the big geographical ideas: scale. Inevitably, all three geographical scales, city, county, and MSA get interwoven. Tables 9-5 through 9-7 have the most populous list at each scale. Understanding the geographical dimensions of growth around Austin, for example, requires the data for three scales. The actual city is growing, other cities and unincorporated areas of Travis County are growing, and parts of adjacent counties are growing.

Texas MSAs

When we construct the metropolitan areas, the rank order changes a little bit. The Dallas-Fort Worth MSA includes 12 counties and totals 5.5 million residents, which makes it the most populous urban region in

Texas. It also ranks fifth nationally after the Philadelphia–Camden–Wilmington MSA, and at forty-sixth in the world, its metropolitan population is similar to Baghdad, Iraq. Dallas and Fort Worth were initially combined in 1970 as single "consolidated" metropolitan area, and there was initially some resistance to the idea because both places were historically independent. Despite the civic rivalry, the term Metroplex evolved as a promotional name thereafter. Three other top ten populated cities are in this MSA: Arlington, Plano, and Garland; in addition, Irving, Grand Prairie, Mesquite, and Carrollton each have over 100,000 residents.

Harris County and the city of Houston statistically dominate the state's second largest metropolitan area. Following the current naming convention, the MSA is called the Houston-Baytown-Sugarland, but could just as easily be called the Greater Houston Region. Technically, the second largest place is Pasadena with 144,000 residents. Previously, Galveston was included on the MSA name for historical reasons, but it currently has only 57,000 residents. The combined population of the 10 counties in the MSA recently approached the 5 million mark. Despite Houston being fourth largest American city, the MSA centered on Houston only ranks eighth nationally and fifty-third globally.

The third and fourth largest MSAs in Texas are adjacent to each other. Both San Antonio and Austin are fast growing cities whose peripheries are quickly sprawling together along the IH-35 corridor. If they were considered a single or consolidated MSA, the population would exceed three million. San Antonio MSA is the larger of the two with eight counties totaling slightly fewer than two million inhabitants. Of the total MSA population, most are residents of San Antonio proper. The city of San Antonio recently passed Dallas as the second largest city in Texas making it also the eighth largest in the country. The five counties of the Austin-Round Rock MSA have a million residents, and they have one of the fastest growth rates of any urban area in the state. A complete list of the 25 MSAs is provided on Table 9-7.

Two additional MSAs in Texas have over half a million residents. The El Paso MSA only counts the over 700,000 residents of El Paso County, but a majority of the functional metropolitan area actually lies in Juarez, Chihuahua. Likewise, the McAllen-Edinburg-Pharr MSA only includes Hildago County excluding other counties in the valley or any of the urban area in Tamaulipas. Rounding out the 10 largest MSAs in Texas, the Corpus Christi MSA comes in seventh rank with less than a half million people in three counties. Interesting note about the remaining top ten MSAs: they are immediately adjacent to an even larger MSA. The Beaumont-Port Arthur MSA is next to the Houston metropolitan area along the Upper Coast. Brownsville-Harlingen MSA is downstream from McAllen and is a major part of the Lower Valley. Killeen-Temple MSA is immediately North of the Austin area along the IH-35 corridor.

Urbanization and the concentrations of people and wealth in cities are not even. There are clear geographical patterns to cities in Texas. In relation to the environment, the majority of all Texans live on the Coastal Plain; most of the large cities began with some form of successful agriculture and access to resources. These cities often instigated the transportation routes and financial markets that kept them integrated with regional trade and not just be shaped by these structures. Recognizing the current growth dynamic, some scholars have coined the term *Texas Urban Triangle*. The triangle is anchored by the big three cities Houston, Dallas, and San Antonio that are expanding far beyond their traditional city centers. Some scholars have suggested that the metropolitan areas could eventually merge into a massive urban agglomeration. While it's too early to determine if such predictions will come true, the growth of urban Texas is tremendous as well as having an impact on land uses and landscapes.

Metropolitan Landscapes

Texas cities are more similar to American cities than different. However, there are a few aspects that set cities in Texas as distinct. Perhaps the most experiential dimension is openness. Texas cities feel more visually open and less constraining to movement. Despite being very populous cities, the amount of space is also greater; therefore, population densities of Texas cities tend to be lower than other American cities. The best explanation for this lower density and wider sprawl is that Texas cities experienced their massive growth after the introduction of automobiles. Texas cities were designed to accommodate automobile traffic with

TABLE 9-5 Largest Cities

City	Population	U.S. Rank
Houston	2,099,451	4
San Antonio	1,327,407	7
Dallas	1,197,816	9
Austin	790,390	14
Fort Worth	741,206	16
El Paso	649,121	19
Arlington	365,438	50
Corpus Christi	305,215	60
Plano	259,841	71
Laredo	236,091	81
Lubbock	229,573	84
Garland	226,876	87
Irving	216,290	94
Amarillo	190,695	
Grand Prairie	175,396	
Brownsville	175,023	
Pasadena	149,043	
Mesquite	139,824	
McKinney	131,117	
McAllen	129,877	
Killeen	127,921	
Waco	124,805	
Carrollton	119,097	
Beaumont	118,296	
Abilene	117,063	
Frisco	116,989	
Denton	113,383	
Midland	111,147	
Wichita Falls	104,553	
TOTAL	10,788,944	

Source: U.S. Census Bureau, 2010.

ample roads, parking, and residential garages. In addition, Texas cities tend to have higher percentages of single family homes than other American cities. The townhouses, duplexes, and high rise apartments that dominate some American cities, especially in the Northeast, are much less common in Texas cities. From a land use perspective, a million "urban" Texans live on much more land than a million New Yorkers, and they are more likely to own their own house with property and a dedicated spot to park their truck.

TABLE 9-6 Largest Counties, 2010

County	Population	Percent Texas	Cumulative
Harris	4,092,459	16.3	16.3
Dallas	2,368,139	9.4	25.7
Tarrant	1,809,034	7.2	32.9
Bexar	1,714,773	6.8	39.7
Travis	1,024,266	4.1	43.8
El Paso	800,647	3.2	47.0
Collin	782,341	3.1	50.1
Hidalgo	774,769	3.1	53.2
Denton	662,614	2.6	55.8
Fort Bend	585,354	2.3	58.1

Sources: U.S. Census Bureau, 2010.

The automobile propelled another feature of development in the South: the strip. Already a common form of spreading out the business activities along a major road, it becomes even more dramatic with the construction of freeways. An early decision by state officials to build parallel running roads adjacent to new freeways means that the sides of freeways would be zoned and eventually developed for commercial purposes. Besides the businesses and parking lots, commercial advertisement proliferated along the sides of these new strips. Billboards became a common landscape feature as one drove toward a downtown on Texas freeways.

FIGURE 9-8 Downtown Houston Skyline
Central Business District from Memorial Park. Photo courtesy of Texas Department of Transportation.

TABLE 9-7 Metropolitan Areas

MSA	Counties	Population
Dallas—Fort Worth	12	6,371,773
Houston—Baytown—Sugarland	10	5,946,800
San Antonio	8	2,142,508
Austin—Round Rock	5	1,716,289
El Paso	1	800,647
McAllen—Edinburg—Pharr	1	774,769
Corpus Cristi	3	428,185
Beaumont—Port Arthur	3	388,745
Brownsville—Harlington	1	406,220
Killeen—Temple—Fort Hood	3	405,300
Lubbock	2	284,890
Amarillo	4	249,881
Waco	1	234,906
Laredo	1	250,304
Longview	3	214,369
College Station—Bryan	3	228,660
Tyler	1	209,714
Abilene	3	165,252
Wichita Falls	3	151,306
Texarkana*	2	92,565
Odessa	1	137,130
Midland	1	136,872
Sherman—Denison	1	120,877
Victoria	3	115,384
San Angelo	2	111,823
TOTAL	77	22,085,169

Sources: U.S. Census Bureau, 2010.
*one country is in Arkansas

The Central Business Districts (CBD) or downtowns of Texas cities have experienced their ups and downs like other American cities. Currently, most CBDs are experiencing some form of redevelopment as well as new skyscraper construction. Some of the redevelopment is directly linked to new facilities for professional sports teams. In fact, very interesting architectural elements are being applied to new buildings such as skyscrapers and sports complexes in Texas. While the downtown skylines are evolving, concurrent street level changes are making Texas cities more attractive for single people and childless couples to live in the city center. These changes include entertainment, shopping, and recreational districts, which also make visiting cities a positive experience.

The emergence of satellite centers around major American cities is also present in Texas. Sometimes referred to as edge cities, these places away from the traditional city centers include tall business buildings, shopping centers, and hotel conference centers. The Galleria in Houston is one such well-known example.

FIGURE 9-9 Austin with Capitol Complex
Oblique image was taken from the Goodyear Blimp, America. Photo courtesy of Texas Department of Transportation.

FIGURE 9-10 Houston Ship Channel, TX 146 Bridge
Modern bridge connecting Pasadena and Baytown. Photo courtesy of Texas Department of Transportation.

FIGURE 9-11 Central Business District, Houston
Urban redevelopment and new light rail station in downtown. Photo by Author.

North of Houston, the Woodlands is another edge city that combines a residential uniqueness dimension. Most of the edge city development in the Metroplex is on the north side of Dallas. These edge cities symbolically reinforce the relative decline of downtowns, but they also demonstrate the continued strength of cities as premier social spaces.

Urban geography is very complex because of the multitude of forces influencing cities such as economic, political, social, and cultural processes. New capital and new aesthetics are competing for a limited amount of opportunity to influence landscapes and profit from popular trends. Changes to the urban

FIGURE 9-12 Mexican marketplace (Mercado) in San Antonio
The Mercado has its own flavor and style, downtown San Antonio, Bexar County. Photo by Author.

geography must work around the preexisting structures and organizations that initially created these cities. As urban Texas absorbs population growth, these big cities are likely to grow higher and outwards. Hopefully, transportation and residential construction as well as support for quality of life can keep pace without major degradations to the environment. In all likelihood, sprawl will be with us for a while as the automobile continues to drive the city.

"Democracy Under Constant Repair"

FIGURE 10-1 State Capital Building with Scaffolding
Austin, Travis County. Photo by Author.

10
Social Structures

Texas is a growing and increasingly complex society. Like all developed societies, it has a large bureaucracy that governs and makes possible all the collective desires of the population. That doesn't mean there is a single vision for what and how government should or could or will work. The notion of a social contract and how it works is under constant self-examination, and many people do not see it or acknowledge it. The democratic process legitimates the social contract albeit in tandem with economic forces like budgets and competition. Some of the self-regulating structures that states prominently construct are medical, legal, educational, and criminal; these can be seen in the contemporary landscape as hospitals, courthouses, schools, and prisons. Each of these respectively have a standard of certification, for example, to be a teacher or lawyer. Texas, like the other 49 states, decides what an adequate education is, what a bar exam includes, what a capital crime is, and who can be a licensed tattoo artist, masseuse, barber, etc. All the while, these standards and definitions are defined in a national scale discussion and even global scale at times. Two prominent dimensions to the idea of social structures are the political and economic realms, or political-economy as a broader idea.

Texas' political economy is part of a larger national and even larger global system. While being integrated with the world, Texas tends to dominate its immediate region because of its large population. Before discussing some examples, the term *political economy* needs to be clarified. On the surface, it may appear as a convenient way to condense two topics into one chapter. Most people consider politics and economics to be two separate fields of study, yet we know that they strongly influence each other. Examples include labor regulations deriving from government bureaucrats and a senator's reelection fortunes riding with unemployment rates. Average Americans know that international commerce and immigration policy, which is being debated by their elected representatives, has both dimensions. Perhaps the most significant example is the neo-liberal initiatives that recent American Presidents have strongly promoted. Free trade policies combine both economic and political aspects; free trade is as much a political openness as an economic prerogative.

A certain group of scholars have been reconnecting the political and economic through their research. Essentially, both are elements of modern power. By any measure, America and Texas have power. American political power begins with the legitimacy that democracy provides to the government because the people elect their leaders. Yet political power goes beyond governance type, America maintains the most expensive military force that has the ability to project power globally. In addition, America exhibits an *exceptionalism* that manifests itself in both foreign relations and popular culture; exceptionalism refers to the extra territorial ways that we impose our norms and legalities on areas outside the United States. American economic power begins with the idea of free market capitalism. The size of the American economy is a source of power that influences what other economies can do. In particular, the standardization of trade and investment rules that dictate how and when capital can flow. In this chapter, we discuss only a few aspects of the political geography and economic geography of Texas. With political geography, we look to the electoral patterns and district-making process. For economic geography, we explore development and infrastructure.

POLITICAL

Political geography is the study of power and territoriality over the Earth's surface. Perhaps the most intriguing aspect is that people have compartmentalized the world into approximately 200 distinct political units that we know as territorial states. For many this list of states or countries and their portrayal on maps is the way we think about the world. Mistakenly, we ascribe the planet's cultural diversity into a notion of nation-state. Typically, a majority of people in a state think of themselves as an ethno-cultural group with common ancestry which is translated into a nation. Therefore, they are entitled to self-determination and some form of statehood. The cultural reality is different with over 5,000 languages representing a better approximation of ethnic identities than the number of states. For these smaller and often minority groups, regional autonomy and/or official recognition is about all they can hope for in the contemporary world system.

The political geography of the United States ranges across the spectrum of geographic scale. At the global scale, America plays a significant role in international affairs and multinational institutions. It is at this scale, the term super-power refers to the United States. At the territorial-state scale, the United States is one of only a few with a federal system that tries to balance power between the national and state governments. American democracy was already functioning when Texas formally joined, and Texas immediately became part of the sectionalism that led to Southern Secession. After the Civil War, Texas typically followed the Southern political model that was staunchly Democratic and eventually transitioned toward the Republican Party. At the local scale, politics is more complicated with numerous counties and many individual precincts voting contrary to the prevailing trends. One of the more telling maps of local interests and differing opinions is the results of the 1861 Secession vote. Overall, those who voted decided to back Secession, but the county by county results show a regional pattern that mirrors the relative importance of slavery.

Electoral Behavior

Voting is a behavior that can be observed for spatial patterns and temporal trends. In some democracies, one can actually watch voting behavior with community assembly and town-hall meetings. More typical in American democracy is the closed ballot box. Because the ballot box vote is considered private, we never really know how any one individual votes. Eventually all the individual votes must be tabulated, so a total vote count can be released. These results are available at different geographical scales. For example, an election for governor reveals not only how the state voted but how each county voted. At the local scale, the votes are publicly counted at every voting precinct in the county before being certified by the top county election official. It is at the precinct level that voting behavior can be correlated with other socioeconomic data such as race, income, and education.

The record for early Texas elections does survive such as the previously mentioned Secession vote. Even earlier, the election of Sam Houston as the first President of the Republic in 1836 is known. Most of the candidates for those earliest elections did not run with party affiliations. After Texas joined the United States, elections became more partisan coinciding with the political trends of Southern sectionalism. Democratic candidates won nearly all statewide races, so slight ideological differences between liberal and conservative Democrats served as the major political cleavage. Similar to the 1861 Secession vote, the primary ideological split fell along the Upper and Lower Southerner migration pattern. The voting pattern appears as a southeast-northwest phenomenon with a line between San Antonio and Texarkana serving as an approximate boundary. Lower Southerners developed a voting behavior that favored more liberal Democrats, and Upper Southerners responded by voting for more conservative Democrats.

Terry Jordan published the political regions map (Figure 10-2) in the 1980s, so he only saw the initial rise of Republican suburbs. At the time, the electoral record supported his eight political regions. In addition to Upper and Lower Southerners, Lower Midwesterners who migrated to the northern Panhandle also diffused political behavior, but in their case, they introduced an allegiance to the Republican Party. Likewise, the German-Texan political region strongly supported Republican candidates, and they were the only pro-Union constituency in 1861 Texas. Both of these traditional Republican regions did occasionally vote for conservative Democrats as a pragmatic manner. In general, many of the Direct European groups tended to vote for conservative candidates while avoiding populist, racist, and third party campaigns. One exception to this rule was the unprecedented independent presidential run of the former President, Teddy Roosevelt.

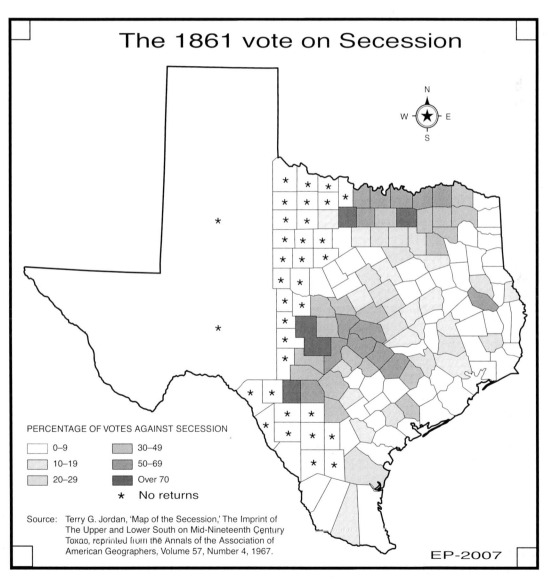

FIGURE 10-2 1861 Secession Vote

In South Texas a unique form of patronage politics developed between land owners and farm and ranch workers. Initially called Patrón after the Spanish land owners; Anglos eventually held the important positions with Hispanic workforces. The Patrón system delivers votes as a block after the boss and workers negotiate ahead of time. Along the border another region with relatively strong support for Democratic candidates formed. In the 1970s, the heavily Hispanic region voted for a short lived party called *Raza Unida* that had played an important role in winning Hispanic representation on city and county governments. East Texas also had complicated aspects due to its black/white composition. Immediately after the Civil War, African-Americans helped vote in Republican office holders around the state. For the most part, the Black population was disenfranchised between Reconstruction and the 1960s Civil Rights Movement. The trend since then is strong support for liberal Democrats. The White population of East Texas also voted for liberal Democrats, but they occasionally voted for populist and racist candidates as well as third parties. For example, East Texas counties provided most of the votes for Strom Thurmond's 1948 and George Wallace's 1968 presidential runs. North Texas and West Texas were the backbone for conservative Democratic candidates. In West Texas, the emergence of an oil boom brought in speculators that changed the political dynamic. Jordan called the region the Permian Basin Republican, which preceded the transformation of conservative Democrat to Republican in other parts of the state. The most important transformation was the growth of suburban areas in Texas that are closely identified with the rise of the Republican Party. While not a large-sized territory, the suburbs have fast growing populations. As the suburbs strongly voted for Republicans, the statewide balance tilted in their favor. On a national scale, Texas cities stand out as leaning toward Republican, whereas most big American cities are strongly Democratic.

The transition from a staunchly Democratic voting state into a solid Republican voting state is very interesting to scholars. This monumental transition in American politics occurred earlier in other Southern states, and in Texas it took hold much more recently. Even though Texans had voted for Republican presidential candidates as early as 1928 (Hoover) and elected a Republican Senator in 1961 (Tower), Democrats controlled the state legislature until the 2000s. One factor in the transition is that conservative Democrats became conservative Republics, which means that the change was more a switch of party than an ideological change. As the national Republican Party developed a conservative persona, conservatives in the South began to identify with them instead of the Democratic Party. We can see from the selected elections in Figure 10-3 that many counties in Texas went from supporting Democrat candidates to Republican candidates. On the national scene, Texas has become the largest reliable Republican state in Presidential politics. In addition to having 34 electoral college votes, three recent Presidents have come from Texas. Lyndon B. Johnson was the thirty-sixth president of the United States; George Bush Senior and Junior were the forty-first and forty-third presidents respectively.

Political Districts

Political districts are another type of geographical region with area, people, and boundaries. Political regions are unique because they are legitimated through elections which reinforce their importance. They often coincide with our social units such as cities, counties, and states that provide our basic welfare, education, and security functions. The political leaders of these units that are elected by the inhabitants are typically also residents of the region. However, not all political districts are the same. There are two generalized categories of political districts: electoral and administrative.

Electoral

Electoral districts differ from administrative districts because the exact boundaries change. The districts change because they must conform to the constitutional requirement of one person one vote. Every decade after the official census counts, electoral districts are redrawn to reflect changes in the population. Most states including Texas rely on their legislatures to redraw and approve the new districts. This means that politicians might actually be drawing their own electoral districts, but it definitely puts the process into the political realm. Drawing boundaries for political advantage is one of the occasional results; when the boundaries

FIGURE 10-3 Texas Political Regions

appear to be overly manipulated it is commonly known as Gerrymandering. Increasingly, these district making decisions are being brought before the judicial branch, which has produced district maps for elections.

Texas recently had a high profile redistricting battle that stretched out beyond the typical year after the census. The 2000 Census was released in 2001, and the redrawing of boundaries was to be in effect for the 2002 election cycle. The legislature, with its partisan split, was unable to agree on a new map, so the redistricting ended up in a federal court. The court drew the boundaries for the 2002 elections. Republicans won enough seats to control the state legislature, and they proceeded to redraw the boundaries for the 2004 elections. Democrats tried to stop the redrawing of districts in mid-decade by denying quorum and initiating legal challenges, but to no avail. Besides drawing their own legislative district boundaries, the redistricting helped change the composition of Texas' delegation to the U.S. Congress.

220 Part Four Modern Human Geographies of Texas

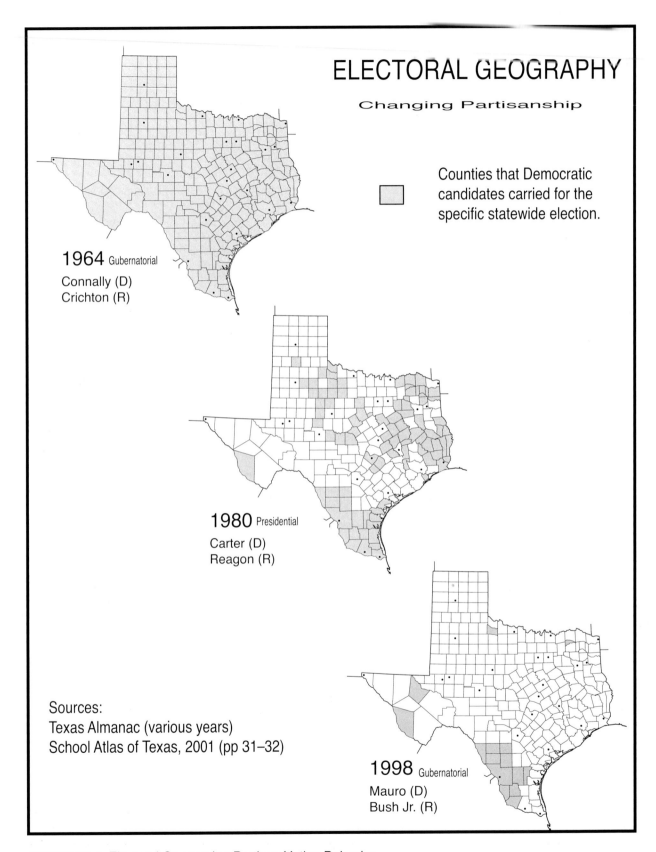

FIGURE 10-4 Electoral Geography: Partisan Voting Behavior

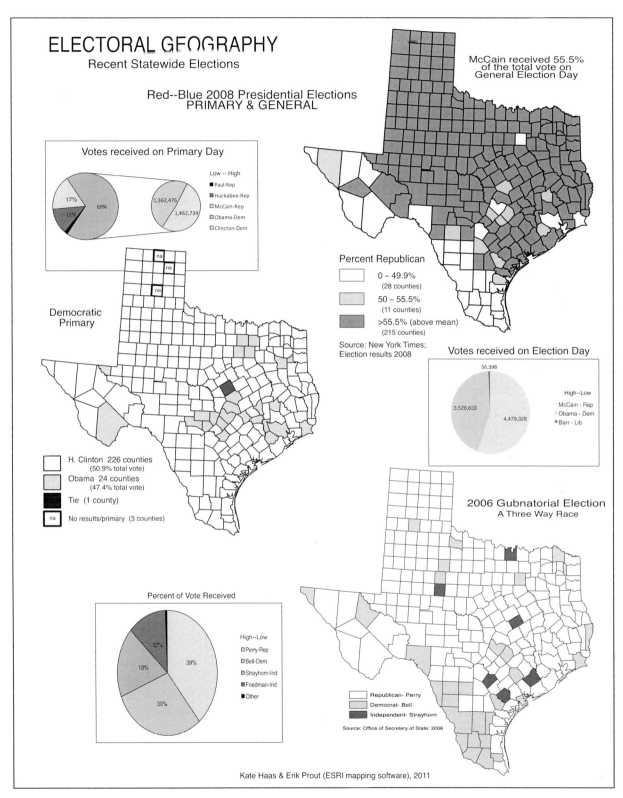

FIGURE 10-5 Electoral Geography; Recent Statewide Elections

TABLE 10-1 Texas Congressional Delegations

Census Year	House
(1845)	2
1850	2
1860	4
1870	6
1880	11
1890	13
1900	16
1910	18
1920	18
1930	21
1940	21
1950	22
1960	23
1970	24
1980	27
1990	30
2000	32
2010	36

Source: U.S. Census Bureau.

The list of electoral political districts begins with the State Legislature. The 31 state senator districts and 150 state house districts must be equally apportioned after every census. In addition, the State Board of Education is elected with 15 equally populated districts. The redrawing of districts for the U.S. House of Representatives changes every decade as well, but the reapportionment of representatives has a beneficial outcome for Texas. Because Texas has grown faster than the national average, the size of its congressional delegation has increased. Currently, the state of Texas is apportioned 32 house seats. Therefore, there are 32 equally populated congressional districts in Texas based on the 2000 Census. As the second most populous state, Texas has the second largest congressional delegation.

Administrative

Administrative districts have their authority vested in them by the state government, and they have no requirement to be equal in size or population. The most prominent administrative units are cities and counties, which vary tremendously in terms of population. The 254 counties of Texas initially derived from Mexican municipalities and evolved with human settlement. In addition to an East to West pattern associated with frontier expansion, many counties are subdivisions of previously formed counties. Texas followed a southern model of relatively compact counties that allowed every resident in the county the opportunity of easy travel to the county seat. On the rare occasion, county seats moved to a more central location. Since Kenedy County was formed in 1919, the county structure has remained stable with minimal changes. Counties perform basic functions for the state such as policing, tax collecting, and administering welfare programs. Elections for officials like sheriff, tax assessor, and county judge are conducted with the county as a single electoral district. However, if a county has a board of supervisors or governing council, those electoral districts have to be equally populated.

For most Texans, incorporated cities are the immediate local government. There are over 1200 incorporated cities in Texas, and each follows its own spatial pattern. The very big cities have very complex governments with thousands of civil servants and trained employees doing every imaginable task. At the other end of the spectrum, many small cities have only a handful of paid staff, and they contract with the county for essentials such as policing. The most common function associated with cities in Texas is education that provides a level of local control over the public schools.

Due to its size, the State of Texas has numerous administrative districts or branches because there is a need to provide the proper scale of efficiency. Examples include state trooper offices and agricultural cooperatives, but these districts do not require elections. Policing by its very nature is extremely geographical because police resources have to be deployed where necessary and are constrained by distance. It wouldn't be feasible to have a single centralized police department to dispatch officers for basic patrol duties. Special purpose regions such as water conservation districts follow the physical geography of water resources. The judicial branch in Texas reflects both an attempt at efficiency of scale and local demand and support for additional branches. The result is 425 state district courts and 80 courts of appeals. In addition, there are 485 county court judges, 828 justices of the peace, and 1371 municipal courts judges. In those instances of voter participation by direct elections or conformation of appointments, the shape and size of the electoral unit conforms to the judicial district.

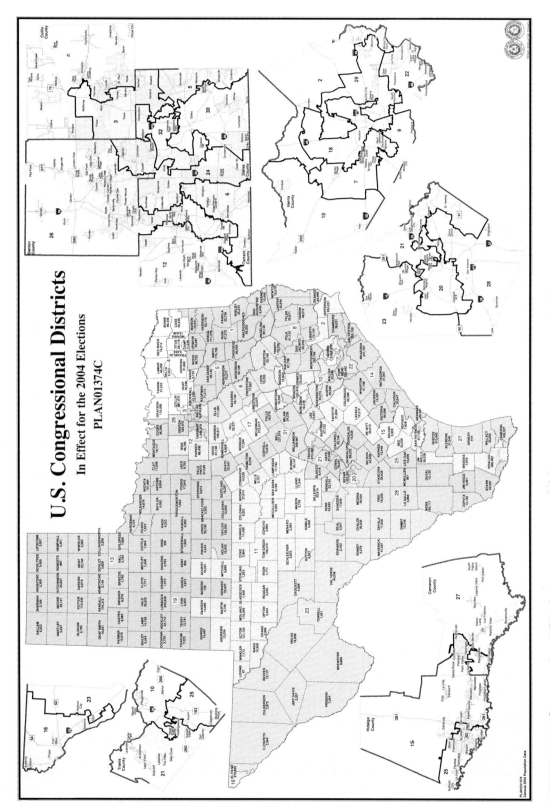

FIGURE 10-6 Congressional Districts
Map courtesy of US Census Bureau.

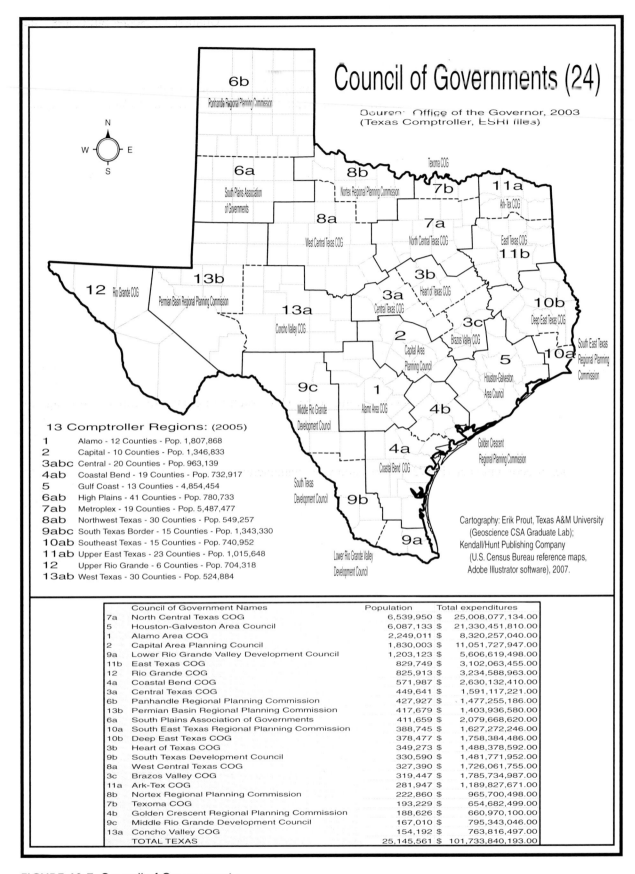

FIGURE 10-7 Council of Governments
Map courtesy of State Comptroller

The state of Texas has experimented with other forms of regional government. The most developed regional governance structure is the Council of Governments (CoGs), which are groups of contiguous counties. These councils perform planning and administer programs that address regional environmental and social programs such as water quality, garbage disposal, and mass transit, which by their nature do not exist in only one county. The individual councils have adopted a wide range of names that include many promotional names. For example, Heart of Texas, Golden Triangle, Winter Gardens, and Big Country are promotional in nature.

The political geography of Texas reveals that the initial migration to the state still influences how the current citizens vote. While there has been a major transition between political parties, the conservative ideology has been a more resilient feature. Texas' role in the overall American political scene has increased with Presidential campaigns, but it also exposes the state to more outside scrutiny and attention. We now turn to the economic geography of Texas.

ECONOMIC

Economic geography is the study of the spatial dimension to production and consumption as well as the global differentiation of well-being and livelihood systems. At a most basic level, economic geography asks what do people do on a daily basis? And then economic geographers can classify different types and categories of work. For example, *primary* economic activities such as agriculture, forestry, mining, and fishing secure resources for the population as a whole. Historically, most humans were engaged in primary activities especially with food production, and the majority of those living in the poorer regions of the world are still engaged in subsistence agriculture.

Secondary economic activities are associated with manufacturing. Simple manufacturing includes the processing of natural resources into a higher value commodity; for example, the making of furniture from forest products. Technologically complex manufacturing such as the automobile industry has multiple resource chains that become constrained to a small number of production locations. The labor force of industrialized states in Western Europe and North America transitioned from primary to secondary, and the transition to manufacturing coincided with urbanization.

FIGURE 10-8 Election Signs
Partisan supporter of Republican candidates. Photo by Author.

FIGURE 10-9 Election Signs
County Democratic Party Headquarters. Photo by Author.

These advanced economies are now experiencing a postindustrial transition, which is quite dramatic for places that had over specialized. In advanced economies, the transition to *tertiary* activities such as services is quite well established. The expression that America has a service economy refers to the workforce dominance of services as well as the importance of consumers and their spending to the overall economic well being. Over half of the workforce in Texas is employed in the service sector in one form or another. Service sector employment ranges from high order financial planning to lower order stereotyped burger flipping.

Development

One of the truisms of economic geography is that economic development is uneven. The uneven or unequal levels of development are observed at various geographical scales. Globally, some states have larger economies and/or more economic diversification than other states. In addition, states have different levels of technology and access to capital investment. The result is that there are a few very powerful economic actors or states and numerous powerless actors. An obvious example of this is the annual meeting of the G8, which is the summit of the heads of state of the eight largest economies; the G8 is widely seen as managing the world economy. The United States plays an enormous role in shaping the global economy because it allows other economies access to its markets. Other economies need access to the large American marketplace to grow, and they become somewhat dependent on the social stability. In return the United States forcefully advocates reciprocal access to their markets which creates a nearly universal global marketplace. Many other states are simply navigating through this system. Occasionally small states such as Switzerland and Singapore carve out a niche that allows them to be successful global players.

Locally, individuals also vary by how much access to technology and information they have. More significantly, individuals in a society have different levels of wealth and influence on the economic system. In both developed and underdeveloped societies, there are those who have more than the average person. The developed societies tend to have a more thorough welfare system that asks those with the most to pay more for helping those at the bottom with a progressive tax structure. Nevertheless, rich-poor distinctions remain. One feature of the developed world is the existence of a large middle class. In the underdeveloped world, there is a more pronounced have-have not divide with a small or nonexistent middle class. The rich have

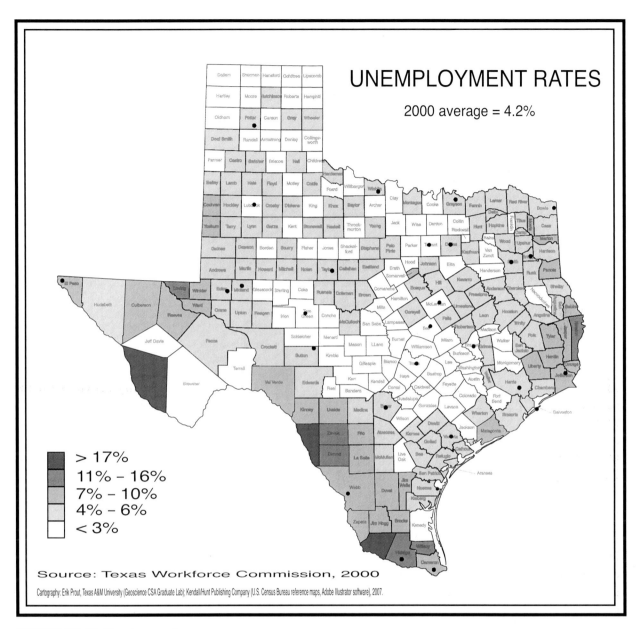

FIGURE 10-10 Employment

the ability to interact with the rest of the world and avoid most of the grinding realities of poverty in their society. The poor are basically cut-off from the ability to access and obtain any benefits from the global system of trade and communication.

Globalization is often defined as the increasing interconnections between places on the planet. The evolution of the trade and communication networks that physically put us into contact with other people is part of the phenomenon known as globalization. Economic forces propel globalization more than any other, and globalization is justified by the logic of capitalism. While globalization has political, social, and cultural dimensions, the economic imperative to open markets has commercially linked the world together. At the global scale, the differences between the well to do in developed societies is starkly better than the disadvantaged in the underdeveloped societies. Places like Texas are situated in the developed

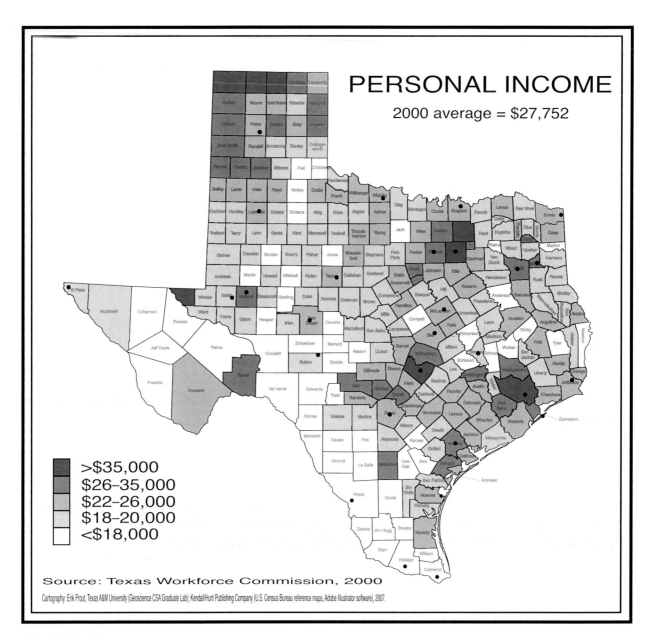

FIGURE 10-11 Income

world, which is capable of interacting with and profiting from the globalization of markets. However, measures of economic conditions vary in Texas like they do in other societies.

Quality of Life

Numerous statistic indicators are useful in discussing the economic well-being of a society. The most common statistics are employment and income related because they are easy for everyone to understand. Employment can be a simple count of the total number of paid positions companies have. For example, the current employment total is approximately eleven million. Interestingly, over half of the state's residents do not work for money, which makes more sense when we remember that the younger than 18 and older than 65 segments are nearly 40 percent of the population. Changes in employment

TABLE 10-2 Economic Indicators

Economic Indicators	Texas (USD)	US
Total Gross Product	1223.511	14093.31 millions
Per capita production	*178,700*	
Labor Force	12,317,200	
Employed	11,270,238	
Growth of employment	3.20%	1.90%
Unemployed	1,046,962	
Unemployment rate	8.50%	9.10%
Total Personal Income	993 billion	
Per capita Income	*39,493*	*40,584*
Personal income growth	7.80%	5.20%

Sources: Wells Fargo; Texas State Comptroller of Public Accounts.

especially the growth rate are considered important indicators of the overall economy. Often employment is measured as unemployment. Unemployment rates are the percentage of the job wanting population that currently does not have employment. The official unemployment figures usually report only those who register for unemployment benefits and not those seeking first jobs and long term unemployed not receiving benefits. Texas has an unemployment rate less than 5 percent which is slightly above the national rate.

Income is the other popular statistic. Total income or total wealth produced would work for understanding the economy as a whole. Typically, income is reported as an average. Median income is one such method that reports the actual income of the individual in the middle of the income rankings. Just as common is the per capita income that is calculated by dividing the total income by the number of people. The current per capita income in Texas is around 28,000 dollars. Texas trails the national average by nearly 2,000 dollars per person. The largest metropolitan areas in Texas all have higher per capita incomes than the state as a whole, and their per capita incomes are above the national number.

Other statistical indicators include measures of economic strength as well as measures of well being. Productivity is commonly reported with economic strength because it one of those pieces of information that investors are very interested in knowing along with work force wages and education. Well being is a more abstract concept. Things like nutrition, infant mortality, and life expectancy are often strongly related to economic conditions, but they do not directly measure happiness and satisfaction. Increasingly, health care issues are seen by many as directly affecting the overall economic strength as well as individual well being, and national politicians tend to address this issue.

Economic Regions

The regional economy is dominated by Texas. Although "regional economy" is undefined, it is clear that Texas plays a large role in the economic activities of our neighboring states. With its large population, port infrastructure, and oil industry expertise, Texas dominates the economic interactions in the western Gulf of Mexico. The Houston-Galveston port complex successfully competes with New Orleans, which has the natural gateway to the American heartland. With its large population and the cultural and transportation links to Mexico, Texas dominates the international trade along the eastern half of the U.S.-Mexico border. Texas is posed to take advantage of the emerging continental trade pacts that are integrating the United States with its hemispheric neighbors both in terms of location and business acumen. The only way to quantify the regional economy is to utilize complex spatial models, but it is apparent from economic

FIGURE 10-12 Economic Regions

measures that Texas is economically powerful. Those places that are economically integrated such as the larger cities and metropolitan areas are generally better off in terms of wealth and employment figures.

The statistical anomaly is along the Mexican border which has significantly depressed economic indicators despite being very integrated. For example, border counties tend to have lower per capita incomes and higher unemployment rates than the rest of the state. Overall, Texas' economy reveals a strong regionalism with the U.S.-Mexico border. The unique relative location of being along this border benefits Texas because the traditional movement of goods and services between central Mexico and the eastern United States must pass through the state. The changing circumstances that NAFTA brought about created new manufacturing in northern Mexico called *Maquiladoras*, which needed to be connected to the American economy. In addition, Texas has become a central node in the movement of people across the international border. In many regards, Texas had always been a human pathway

TABLE 10-3 Economic Regions

Region	Per capita income	Unemployment rate	Poverty rate
Alamo	36,263	7.3%	16.5%
Capital	37,544	7.1%	13.6%
Central Texas	33,984	7.3%	20.1%
Gulf Coast	46,179	8.5%	15.3%
High Plains	33,898	6.1%	17.7%
Metroplex	41,318	8.3%	14.3%
Northwest Texas	35,403	7.0%	16.6%
South Texas Border	25,032	10.4%	30.4%
Southeast Texas	34,097	9.9%	18.9%
Upper East Texas	33,894	8.3%	17.4%
Upper Rio Grande	29,390	9.5%	23.6%
West Texas	38,183	6.6%	15.9%

Source: Texas EDGE Data Center, Comptroller Economic Fact Sheet, 2009.

between Mesoamerica and the American South. South Texas as a sub-region of the state is most characteristic of this cross-border and transnational phenomenon. As a migration entry point, South Texas is statistically different because of the demographic characteristics of migrants who tend to be younger and male for example.

However, one of those characteristics is lower economic development and higher poverty rates than the rest of the state. The "border" has become a political issue at both the national and state levels mainly because of illegal migration, but Texas has tried to address the economic disparities. Texas takes a common approach that minimizes the abject poverty and to insure basic levels of housing, health care, and education. Some of the programs are unique to the borderlands, whereas others are open to any qualified county in Texas. The state also operates an economic enterprise zone program that provides tax credits to encourage economic investment in specific areas. The enterprise zone applies to both poor counties and poor cities; for example, a rich county like Dallas has designated zones to help development in the poorer neighborhoods. Despite shoring up the welfare net and offering tax exemptions, the economic development of the borderlands is not resolved by any means.

When forced to divide the state into economic regions, the major metropolitan areas will stand out as cores. One can think of these cities as engines of growth with a concentration of wealth and employment, and they are surrounded by a suburban hinterland of bedroom communities. In numerous ways, the web of economic interactions is strung most thickly between the three points of the urban triangle, and the rest of the state is thinly connected to the triangle or to the national and international movements of goods, services, and people. Therefore, the economic regions associated with the large metropolitan areas will have the highest employment as well as generally consistent growth indicators.

From a cartographic perspective, the delineation of economic regions must also address local sensitivities and follow a statewide generalization scheme. If we didn't most every county in Texas would appear to be a peripheral part of Houston, Dallas, and so on. Instead, economic regions appear as logical compact units similar to the climatic divisions that are utilized for reporting weather. Recently, the state changed its reportable economic regions from ten to thirteen. From the Central Texas region, they delineated a new Capital region that focuses just on the Austin metropolitan area. The state divided the South Texas region into three more useful units because the prosperity around San Antonio was masking the development disparities of the border counties despite both experiencing growth. In addition to an Alamo (San Antonio metropolitan area) and South Texas Border, the state created a Coastal Bend region that includes Corpus Christi and Victoria.

Infrastructure

Economic transactions rely on the ability of people to make contact, deliver products, and have mechanisms for measuring value. Infrastructure is the name scholars give for all of the underlying support for the economy. Yet, infrastructure underpins society as a whole as it supports the more fundamental ability of humans to live in dense settlements that characterizes modern economies. Economic and social developments are futile without the infrastructure to make things happen. Historically, roads, marketplaces, and standardized units of measurements and currency were essential types of infrastructure. For example, when one buys a pound of flour or a gallon of milk, what exactly are a pound, a gallon, and for that matter a dollar? The U.S. Constitution explicitly gives that authority to Congress. It turns out roads and many other traditional types of infrastructure remain very important. Currently, telecommunications, computers, and access to information are basic elements of economic progress and development strategies, so these are conceptualized as types of infrastructure. In the following section, we look at some of the various infrastructures and elaborate on Texas examples.

Roads

Roads are still premier forms of economic infrastructure because we use automobiles for our personal mobility and roads define the pathways between home, work, and leisure. Road infrastructure is both extensive and intensive in Texas. It is extensive because it goes nearly everywhere, and it is intensive because it overlaps and becomes quite dense in settled areas. Roads have evolved into high speed highways and controlled access freeways. Texas has more miles of road than any other state. There are over 77,000 miles of state-maintained roads in Texas; the state maintains four different networks: interstate highways, U.S. highways, state highways, and the FM (farm to market)/RM (ranch to market) roads.

Yet, the state maintained roads account for just over a quarter of the total roadway with cities, counties, and even individuals maintaining the majority of the state's roads. Some scholars estimate that automobile related infrastructure accounts for nearly half of the land use in cities. The space devoted to automobiles would include roads, highways, driveways, garages, and parking lots, as well as businesses such as retail dealers, gas stations, and repair shops.

Railroads

Texas also has more miles of railroad track than any other state. Railroads were a crucial piece of transportation infrastructure that coincided with the opening up of the western frontier and the rise of industrial power in America. In Texas, the railroad was aggressively introduced by Houston business leaders as a way of linking the rural agricultural areas with the Houston Ship Channel. Therefore, one element of the geographical pattern of railroads in Texas is the interconnectedness of Houston with radial lines focusing on the Galveston Bay ports.

A second element is that the southern branch of the transcontinental railroad network passes through Texas. Many of these East-West lines pass through the Metroplex and eventually bottleneck at El Paso, which is a major decoupling point for this traffic. San Antonio continues to serve as a railroad gateway with Mexico, and they connect to both the Houston and transcontinental networks. The pattern of rail traffic as it leaves the state shows that Texas is strongly connected to the Midwest. While passenger trains and their depots have nearly disappeared from the American transportation scene, railroads continue to be economical ways to move extremely heavy cargo. Freight traffic includes coal, iron ore, grains, automobiles, as well as the ubiquitous cargo containers.

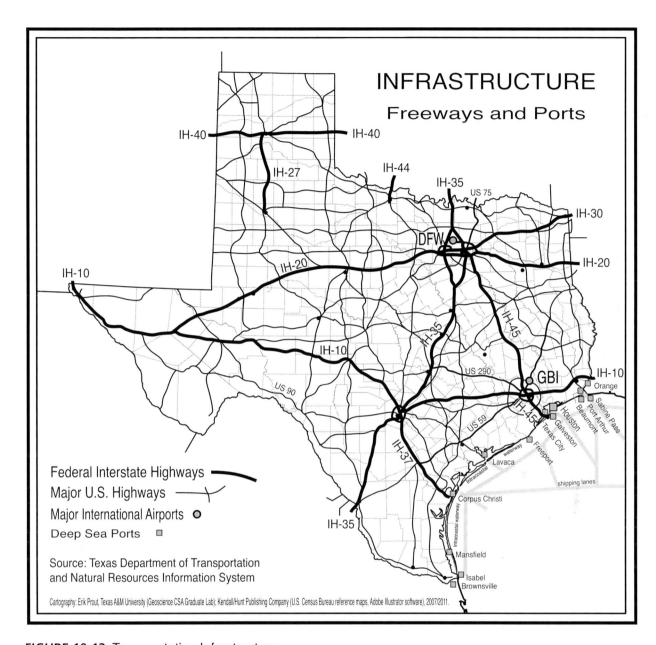

FIGURE 10-13 Transportation Infrastructure

Ports and Airports

Ports are the traditional gateways between Texas and the world economy. Intercontinental trade is limited to ship traffic except for small and valuable items that can be flown by aircraft. Trade by ship has to go through a break of bulk point when the material is loaded and unloaded to land based transportation. The facilities to do this reloading are associated with seaports. Texas has 13 deep seaports that can accommodate ocean going vessels; however, the Houston Port dominates the statistics for all seaport traffic. The term port is also used to describe official entry points into the United States such as crossing stations along the Mexico border.

Interestingly, the word port was combined with the prefix air to label the places where aircraft could load and unload people and cargo. Airports that serve international routes are typically considered points or ports of entry as well. The two largest airports in Texas are the Dallas-Fort Worth (DFW) and George Bush Intercontinental in Houston. Aircraft have increased their role in transporting people and light cargo such as the mail.

Utilities

Utilities are the unglamorous elements of infrastructure that are often treated better if they are unseen. Water is an essential for people, and in contemporary cities much effort and expense is made to provide potable water on demand. Municipalities acquire water from surface or ground sources, treat the water with chlorine and fluoride, and then deliver it to each and every household. The networks of delivery pipes are almost universally buried in the ground, but the above ground storage tanks are often appropriated as community billboards. On the other end of a water delivery system is a waste water collection system. Municipalities must treat sewage before releasing used water back into the environment.

FIGURE 10-14 Water and Electricity Infrastructure, Archer
The Last Picture Show was set and filmed in Archer City. Photo courtesy of Texas Department of Transportation.

Electricity and gas are essential utilities because they power our energy intensive lifestyles. For the most part gas is used for cooking and heating. Electricity can also provide ways to light, cool, and power our gadgets. A national effort was conducted in the 1930s to ensure that electricity networks would reach everywhere; the rural electrification programs are considered by many as one of the federal government's bigger accomplishments. The electricity or more broadly the energy network is quite complex because it requires numerous large scale production sites as well as a universal transmission network. In addition, there are numerous pipelines that move oil and natural gas between sites, and there are coal and uranium supply networks to feed electricity production. Energy eventually became a mixture of cooperatives and big companies that operated under government supervision and regulation.

The invention of the telephone and its almost universal acceptance created a phone infrastructure that grew beyond the simple laying of wires that the earlier telegraph system did. Eventually, phone lines were strung to nearly everywhere with a similar federal program to electricity which helped rural America become wired. Unlike energy and water, phone utilities remained private companies that would be regulated

FIGURE 10-15 Historical Infrastructure
Route 66 near Amarillo; Texaco Station, Dallas Backtracks; Old international bridge at McAllen; Hurricane damage, Lavaca Bay Causeway.
Photos courtesy of Texas Department of Transportation.

by government and occasionally become embroiled in monopoly controversies. In recent decades, many parts of the country were wired with cable television, and even more recently, some places have fiber optic or high capacity wires to accommodate high volumes of data transfer. In all likelihood, phone, television, and internet will become a single piece of infrastructure in the future.

Education and Information

Recently, politicians and scholars have considered education as a type of infrastructure. Once one understands how important basic forms of infrastructure are to the economic well-being, they can be used to describe other aspects of economic strength. For example, a well educated population is directly related to productivity and wealth. Therefore, we can conceive of education as a basic foundation for economic prosperity. As an infrastructure, education includes the networks of school districts, teacher training, and

FIGURE 10-16 Modern Infrastructure: Rail and Road together
The other side of the tracks from the JFR memorial downtown Dallas. Photo by Author.

up-to-date learning environments that include computers. The measure of this infrastructure would be the number of graduates prepared for the demanding realities of the future global workspace.

Another dimension to economic well-being is the access to information. Globally, those economies infused with technology and information are likely to be larger and wealthier. Increasingly, we view information as a foundational component for the overall economy. The information infrastructure would include not only the computer hardware and software but the training associated with accessing, storing, and utilizing data for productive purposes. The connection between information training and education cannot be clearer.

The political-economy of Texas is powerful with numerous indicators of political and economic strength. As Texas grows, we can anticipate more power and influence in the regional, national, and global systems. Nevertheless, the economic disparities and partisan lopsidedness are troubling aspects that always haunt democracies and free market societies. Texas' relative location in the American Southwest along the Mexican border will be a source of dynamism. On one hand, the demographic and economic contribution will spur growth and change. On the other hand, discomfort and even resistance to change may play a spoiler role in future prosperity.

"Curious People at a Curious Art Site"

FIGURE 11-1 Cadillac Ranch, Amarillo
West of Amarillo on IH-10, Potter County. Photo by Author.

11
Leisure and Play

One should not overlook the impact of leisure in our society because of an ignorance of geography. While the geographies of leisure might not be a frequently spoken expression, the spatial dimension to leisure is tremendous. All developed societies have built up a sizable middle class with disposable income, and average citizens can spend time and money in ways they control. Broadly conceived as leisure, it includes shopping at the mall, taking vacations, going to the gym, and pursuing recreation in all its guises. Americans have integrated the automobile into more than just economic realms, and it has become a fundamental part of our leisure. We have fun in our cars, we use cars to go places, and of course we simplified the long-distance road trip. Think about the network of highways and freeways with their multiple places to eat, sleep, and refuel that permits people to travel long and far with a dizzying array of possible routes. All these dimensions to leisure have their own geographies, and, increasingly, they are becoming a more significant visual component of the cultural landscapes in America.

Many of our leisure activities could be compared with play. The road trip becomes an adventure; the lap around the track becomes a race; the cheering on of the home team becomes an intense activity. Play is not just for the young, it is big business with social implications that are worthy of serious research. Numerous studies show that children's play is a form of preparation for adulthood, where role playing is tested and social skills are formulated. For example, games such as house are, in addition to mimicking of one's own household, a first go at gender roles. Generally speaking, playgrounds are sites of interaction for the development of social skills. A short list includes sharing a toy, taking turns at the slide, and joint responsibility for fun on the teeter-totter. While these idealized skills are forming in children, many adults flaunt their childish aspirations. You may remember seeing one of those bumper-stickers that say the one who dies with the most toys, wins. Increasingly, play for both adults and children is intertwined with consumerism. In addition to the direct acquisition of video games, recreational vehicles, and other material items, ordinary people purchase experiences such as cruise ship vacations and luxurious pampering at a day spa. Of course, there is a corresponding dimension of promotional advertising and marketing of leisure that includes very powerful visual representations of the good life. These images of the most desirable places for leisure define our collective desires as well as interject a consumptive landscape norm.

The geographies associated with these activities are numerous, which include widespread land uses and distinct landscapes. At one scale, the designation of parks that have well-defined activities such as camping, horseback riding, and off-road vehicle use can occupy large tracts of land. Big Bend and Big

FIGURE 11-2 Queen Theater awaits restoration
Downtown Bryan Historical District, Brazos County. Photo by Author.

Thicket exemplify this large scale with multiple agendas that prioritizes ecological preservation. At the other scale, individual buildings such as a retail store or a movie theater are ubiquitous elements of the current cultural landscape. However, they reflect everyday pastimes as well as the ordinary time–space dimensions to leisure. In fact, most downtowns in Texas have an old movie theater that may even still be in use. Archer City comes to mind because it was the set for the Oscar-winning movie The Last Picture Show. Since the movie, residents have restored the theater, albeit without a roof, which allows it to be an outdoor setting and an open tourist attraction. While most small downtowns have declined in economic and social importance, shopping and leisure, in general, have increased. The mall has become the preeminent shopping experience in America, and these large complexes tend to be located away from historical downtowns. A new pattern or geography of transportation and parking emerges around these contemporary shopping malls; likewise, a predictable accommodation of cars is found with strip malls along highways and major boulevards.

In this chapter, we explore the multifaceted geographies of leisure, which includes tourism and the role of describing Texas to outsiders. While tourism regions are extremely generalized, they are interwoven into

> **LOVE OF A CAR**
>
> One often hears the expression, "Americans love their cars." For over a century, the automobile has been a part of the American scene—quite literally as an integral part of the cultural landscape. Automobile production is iconic for understanding the modern assembly line and more broadly the economic system that is commonly called Fordist, which was named after Henry Ford's mass production of the Model T. Furthermore, the American automobile industry is symbolic of industrial strength and the rise of the United States into a global superpower. Many social institutions such as labor unions were formed and somewhat defined by the relationship between automobile industry management and auto workers. Furthermore, American politics incorporates this social dichotomy between business and labor as one of the defining differences between the two major political parties. Now that globalization and postindustrial economics dominate, the automobile industry is symbolic of the difficulty of change and restructuring. In addition to the obvious economic impact, the car has helped shape American culture. Geographically, the automobile extends and accelerates human mobility. It gives the individual the opportunity to migrate, travel, and commute alone over long distances. With low energy prices, it allows a sprawled out settlement pattern that will continue to define American cities for the foreseeable future.
>
> A list of examples of how we love our cars includes cruising and racing, which emphasizes individualism and one's personal connection to their cars. There is a general fascination with what is technologically possible in terms of speed and customizing of an automobile that can border on obsession. Perhaps the most significant cultural aspect, which is often overlooked, is love and sex. The car has become a part of dating and courting as well as a complicit aspect of unsanctioned sex. Most think of true dating as something that takes place after one obtains the autonomy of a car; it's in the car that making out at Lover's Lane is conceptualized. The car becomes a means to go somewhere else for secrecy and illicit affairs. The car actually plays a role in the creation of the motel, which, as a word, is the contraction of motor-hotel. Moreover, the motel is associated with sex because of its location along highways far enough away from home to bring a sense of secrecy. Similar to the link between car and sex, which transforms the motel into an American institution, the car itself is a site for sex. In surveys, over 70% of Americans have positively responded that they have had sex in a car. Studies suggest that the automobile is the second most likely place for teenagers to lose their virginity. In conclusion, are we in love with our cars, or are we loving the connection between cars and sex?

more complex economic and demographic realities. Furthermore, tourism regions and their toponyms may play an important role in future, popular understandings of the state. We begin with the scholarly approaches to tourism and a thorough discussion of the official tourist regions. Then, we turn our attention to the more play-oriented activities, and explore some of the possible geographies associated with leisure found in Texas.

TRAVEL AND TOURISM

Scholarly work on travel and tourism is quite extensive, and it includes research on the financial statistics of this phenomenon. The volumes of people traveling for both business and recreational purposes are increasing, which provides many with the opportunity to observe firsthand the world's diverse cultural and physical geographies. As an economic force, travel and tourism in all its forms is becoming one of largest sectors of the world economy. The World Travel and Tourism Council claims the growing value at over six trillion U.S. dollars. In addition, they attribute around 3.6% of the world's GDP to the travel and tourism industry, and they estimate it is closer to 10% when you add the supporting activities. Ordinary Texans can reasonably expect to travel in their lifetimes. They will not only visit places in Texas but destinations in America and the world. Likewise, we can expect visitors from near and far to impact our state and local economies.

Geographical research on travel and tourism tries to understand this flow of people and the impact it makes on specific places. Tourists vary in where they come from and where they want to go. In addition, tourists have multiple motivations that vary from doing nothing at a spa or simply watching a sunset to extreme experiences such as rafting and rock climbing. Both natural and cultural landscapes reflect these driving forces. We can categorize many of these tourist types for Texas: historical, conservation, and diversion. By no means is this a comprehensive list, but it serves as an organizational structure. We begin with the most recognizable icon in the state—the Alamo.

FIGURE 11-3 Tourists at the Alamo, San Antonio
Visitors read the interpretative signs inside the Alamo grounds, Bexar County. Photo by Author.

FIGURE 11-4 Riverwalk, San Antonio
Sidewalk cafes along San Antonio River waterfront (Paseo del Rio), Bexar County. Photo by Author.

Historical Tourism

Occasionally, a historical site becomes the strong focus for tourism. The Alamo serves such a function for San Antonio as the most visited tourist site in Texas. In a way, San Antonio itself with numerous missions including the Alamo, La Villita (original chartered city), and Riverwalk (Paseo del Rio) is thought of as a historical place. Of course, the city lures visitors in with its unique setting and cultural past, and then it capitalizes on a long list of amenities including theme parks. Riverwalk exemplifies both a charming setting and intense services associated with dining and shopping. Currently, San Antonio has an annual visitor count of around 26 million, which is equivalent to the state's population. Big cities like San Antonio have the necessary infrastructure such as hotels, convention space, and an airport that support both business and leisure tourism. Both the Metroplex (Dallas–Fort Worth) and Gulf Coast (Houston) exceed San Antonio's capacity to house visitors, but the tourist literature portrays them primarily as business centers, with leisure and recreational opportunities being secondary. Nevertheless, they and other cities draw on their historical dimensions—most notably with Texas stereotypes such as ranching—albeit in tandem with their modern amenities.

Not to be outdone, small towns around the state display their historical past. Small town Texas has its own charm and sense of community that others would feel very welcome to visit, but most of these places do not have the capacity to house, feed, and entertain large numbers. They contain numerous historical buildings and esthetic landscapes that conform to the stereotypes of Texas such as Southern, Western, and Southwestern. Smaller towns generally benefit by regional-scaled tourism promotion efforts that link them together with other small places. Therefore, many smaller towns are typecasted into themes and categories. Historical tourism as a whole reflects people's desire to understand the past, and it often combines an educational motive with an experience of being in specific places. The educational aspect might be personal understanding as well as intergenerational learning, which seeks to introduce and indoctrinate children with historical facts. The place aspect includes being at the actual site of some historical event or artifact such as the Alamo, but place can also be the setting for making memories. These historical sites become a real place for memory making as well as the literal backdrop for taking photographs that become a memento in itself.

Of all the possible places for historical tourism, I highlight just two particular types in the cultural landscapes of Texas: museums and archaeological sites. They stand out because they overlap with professional efforts to discover new facts and make informed interpretations. Museums are widespread and numerous in Texas. The very large cities have history and science museums with trained curators and tour guides. The Bob Bullock Texas State History Museum in Austin and the Fort Worth Museum of Science and History are good examples. However, the majority of museums in Texas are small in size, more likely to be single themed, and even eclectic. While history museums collect artifacts and professionally interpret the past, Texas has numerous actual sites where historically important activities occurred. The Alamo and San Jacinto Battlefield Site exemplify this monumental reverence towards actual sites, and they attract

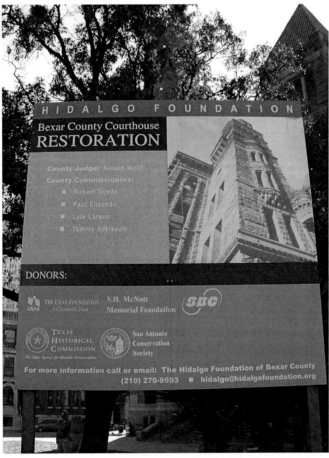

FIGURE 11-5 Historical Tourism and Courthouse Restoration
Bexar County Courthouse, San Antonio, Texas. Photo by Author.

large numbers of tourists. In contrast, most active archaeological sites such as Fort Saint Louis are not open to the public, but places such as Fort Davis are being preserved so that present generations can see what a frontier fort looked like. At the Fort Davis National Historic Site, the footprints and remains of structures are being kept in a quasi-dig setting much like a museum display. Fort Davis is distinct because of the service by Buffalo Soldiers, so an *in situ* discussion of race on the frontier can be facilitated. The state also maintains the surviving Spanish Missions, including their churches and any remaining infrastructure such as irrigation ditches. Quite literally, these sites of culture history are conserved for future generations, and our next category of tourism attempts to do the same for nature and ecology.

Conservation Tourism

Local, state, and federal governments establish policies directed toward environmental conservation, and many people support these goals of maintaining and protecting nature. As related to tourism, environmental conservation provides actual locations for visiting and interacting with the natural world. However, the primary or equally important consideration of environmental conservation is the protection of plants, animals, and ecosystems and not necessarily the comfort of human visitors. When appropriate and legal, human activities that extract resources and provide enjoyment coexist with environmental goals. Geographically, those areas with human exclusion and coexistence are both manageable and mappable. Perhaps the most blatant geographical reality is the enormous tracts of land that have been set aside for conservation. These land set-asides include National Parks, Forests, Grasslands, and Wildlife Refuges operated by the federal government, and the State of Texas has its own State Parks, Historic Sites, and Natural Areas. While having these large set-asides, Texas differs from the American West in terms of not having expansive federal lands managed by the Department of the Interior. This is because the state maintained the public lands from the Republic of Texas and sold much of it to private ownership. A majority of the land set-asides, National Parks and State Forests, for example, are designated multiple-use that allows some recreational activities and resource extraction, which on the surface goes against strict preservation. In addition, those landscapes in dual-use areas reflect a concern for what visitors actually see, so the esthetics of nature becomes another variable of the human–environment relationship.

Most conservation tourism is actually sightseeing or passive admiration, and much of it is dependent on automobiles. Many tourists must rely on their personal automobile to visit natural sites and some even spend more time inside the car than outside. Yet, a definite segment of conservation tourism is defined by experiencing nature in a more primitive way. For example, backpacking off the beaten trail and kayaking

TABLE 11-1 Tourism by Region

Tourism Region	Person-Stays (millions)	Person-Days (millions)	Texan %
Big Bend Country	5.81	13.8	55.8
Gulf Coast	47.0	109.0	64.5
Hill Country	23.1	47.6	76.0
Panhandle Plains	20.6	37.2	72.0
Pineywoods	12.0	21.9	78.3
Prairies and Lakes	60.1	128.2	55.9
South Texas Plains	29.8	70.2	66.0
TOTAL	198.4	427.9	

Sources: Office of the Governor, Economic Development and Tourism.

FIGURE 11-6 Conservation Tourism and Lake Recreation
Rafters on Inks Lake, Burnet County. Photo by Author.

the bayou puts adventurous individuals among the elements. Hunting and fishing are closely related activities, yet these are regulated especially on public lands. The state of Texas collects over a million dollars a year on public hunting and fishing permits. Yet, most of the hunting occurs on private lands because most of the land in Texas is in private hands. Therefore, hunters usually need to make arrangements with landowners ahead of time; for example, hunting leases are very popular in Texas. The state actually collects more revenue, nearly two million dollars, from wildlife management permits. The scale of hunting in Texas can be summarized by the annual harvest of deer that stands between 400,000 and 450,000. The wider economic ramifications of conservation really fall onto the private sector. With many different outdoor activities, numerous private business opportunities exist for things such as equipment, supplies, licenses, and guides. Camping in both improved facilities and primitive sites is part of the interaction with nature that conservation provides. Much of the overnight accommodation is actually provided by private businesses on the periphery of government parks.

Diversion Tourism

Diversion tourism is defined by those activities that transcend one's normal identity and typical routines. These diversions include both the serene meditation of an individual and the collective play of the masses in an amusement park; they range from the sacred spiritual journey to the profane libidinal experience. By its very nature, diversion tourism tends to have explicit cultural landscapes so there is minimal miscommunication between the diversion seeker and the diversion enabler. On one end of the spectrum, intense effort at creating a unique popular experience and/or accommodating large numbers of visitors leads to overly manipulated landscapes. Tourist-oriented places led by Disneyland exemplify how dramatically a landscape can be designed for fun and pleasure. In Texas, Six Flags and Seaworld best exemplify these large-scale, theme-based, manipulated landscapes.

Social critics emphasize the consumerism associated with these festive and playful places along with the excesses of capitalism itself. Theme parks, rose gardens, boardwalks, carnivals, county fairs, and even many new shopping malls are further examples of this trend that deliberatively designs a festive atmosphere for consumers. Without a doubt, developers hope that their investments with these celebratory landscapes will be extremely profitable. On the other end of the spectrum, small-scale diversions, such as religious shrines and tattoo parlors, can be easily observed. However, the illicit and illegal component to tourism can be more difficult to observe and interpret in the cultural landscape. There is no mistaking that geographies of illegal sex and drugs such as prostitution and trafficking exist, but it might be unseen in the landscape unless one is personally involved. Diversion tourism includes a wide range of activities because it reflects the diverse population as well as the variations in personal choice and desire. Now, we turn to the official tourism regions and identify some of the specific places and attractions in each region.

TOURISM REGIONS

The official designation of tourism regions creates a framework for the state to provide significant financial backing for promotion and development activities. By necessity, the state prefers to limit the number of tourist regions and settle on a single scheme. For this reason, there are only seven tourism regions in Texas. A generalization of the state's cultural and ecological diversity into seven geographical regions is immense. One justification state officials generalize so much is that they are trying to keep it simple. Tourists are visiting Texas for leisure purposes, and a simple conception of regions is more likely to be communicated successfully. Most of the tourism regions have only one major place and/or primary attraction. Outwardly, all seven regional names are environmental toponyms using physiographic terms such as plain, prairie, hill, woods, and coast. Two names aggrandize into "country," implying size and character, which is a metaphor for Texas. With so much generalization, each region will have inconsistencies between the dominant stereotype and outlier places with very different realities. The following description of the seven tourist regions locates and tries to highlight their essence as well as some of their inconsistencies.

Piney Woods

The regional name Piney Woods, sometimes contracted as Pineywoods, is an already well-established toponym, and it coincides with the popular understanding of East Texas. As a tourist region, Piney Woods emphasizes the forest environment and recreational opportunities as well as the historical connections to the Confederate South. Big Thicket, with its dense pine and cypress forest, is the prototypical image for tourism but it only occupies a small portion of the Southeast. Lufkin and Nacogdoches compete for top promotional billing with water playgrounds that reservoirs such as Sam Rayburn provide. With its proximity, the Southeast tends to be strongly connected to metropolitan Houston and the Upper Coast. Southeast Texas tourism promotion efforts such as websites are centered in Beaumont. There is a pronounced latitudinal distinction with the Northeast that has more of a post oak landscape and less forestry. Numerous historical places such as Boston, Jefferson, Carthage, and San Augustine that served as gateways for historical overland migration are less prominent in the contemporary landscape. Those places that profited from the East Texas oil fields such as Tyler and Kilgore are much larger. Likewise, Tyler, Longview, and Marshall benefit from the interstate highway system; meanwhile, the highways literally bypass the smaller, left behind places. An exception is Jefferson on Cypress Bayou that has preserved its historic core, and claims to have the highest per capita number of Bed and Breakfast establishments in the state, which makes it a unique getaway destination.

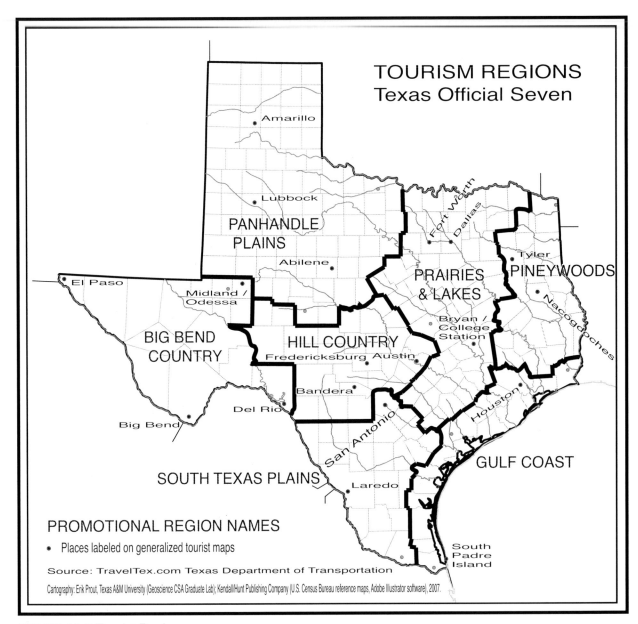

FIGURE 11-7 Tourist Regions

Big Bend Country

This tourist region extends throughout much of the Trans-Pecos and stretches from Big Bend all the way to El Paso. Big Bend Country is commonly represented as a desert environment with natural wonders and Mexican heritage. Big Bend has interesting geological structures and ecological communities, but it does not have much cross-border movement because both sides of the border are extremely remote. The region's largest and most populous place, El Paso, is where the U.S.—Mexico border is extremely busy, and it is ironic that the bustling population of El Paso is somewhat defined by a far-off natural feature. Big Bend National Park is a remote tourist destination with distinct gateways: Marathon, which is more of a

FIGURE 11-8 Historical Square, Nacogdoches
Brick construction and brick pavement of old courthouse square, Nacogdoches County. Photo by Author.

FIGURE 11-9 Nature under Construction
Demonstration Tree Farm in East Texas Timberlands, Newton County. Photo by Author.

cutoff, and Alpine and Marfa that maintain more services and accommodations than the park itself. Meanwhile, the places along the busy interstate highways to the north of Big Bend, such as Fort Stockton, Pecos, and Van Horn support much more traveler infrastructure. Awkwardly attached to the Big Bend Country tourist region is the Permian Basin oil towns of Midland and Odessa, which could easily be part of the adjoining region.

FIGURE 11-10 Big Bend Accommodations
Panther Junction, Big Bend National Park, Brewster County. Photo by Author.

FIGURE 11-11 Highway Business Services
Local and Global compete along US-180 in Lamesa, Dawson County. Photo by Author.

Panhandle Plains

The Panhandle Plains region combines both the high plains including the Llano Estacado and the low plains of the Osage into a single unit. Interestingly, by including both the upper and lower plains, the Caprock Escarpment and places such as Palo Duro Canyon are inside the region instead of a border between regions. Representations of this tourism region typically begin with the stereotyped flat open spaces. Many representations also depict the Panhandle Plains with agricultural images such as ranching and grain harvesting that distinguishes the region from the urban realities of Texas cities. Often overlooked is that this region was settled much later than the rest of the state, so settlement coincided with changes in technology and transportation—specifically the railroad. The modifier Panhandle instantly locates the region on a map, but the tourist region extends far to the Southeast of the geometric Panhandle shape. The region includes five medium-sized Texas cities: Lubbock, Abilene, Amarillo, Wichita Falls, and San Angelo. Each has its own central business district and hinterland; however, much of the current travel and tourism infrastructure can be attributed to the interstate highway network.

South Texas Plains

The largest place and dominant attraction in the South Texas Plains is San Antonio, but only because San Antonio is drawn together with most of South Texas. From the state's perspective, the tourist region stretches southward and southwestward from San Antonio to the Mexican border, but not to the Gulf Coast. The Alamo, which is perhaps the most iconic image in Texas, is used as a symbol for the entire region. It is common for this region to highlight its Mexican culture with scenes from San Antonio's Fiesta. The idea and understanding of Tejano corresponds nicely with this tourist region in South Texas. Interestingly, the adjacent Hill Country counties are separated from San Antonio. As another region portrayed with its Mexican heritage—rightly so, the reality of San Antonio is different, with the German

FIGURE 11-12 Beach Resort Tourism
Corpus Christi Bay and US-181 Bridge, Nueces County. Photo by Author.

FIGURE 11-13 Boca Chica: End of the Road
Boca Chica Village sits downstream from Brownsville near mouth of the Rio Grande; Brazos Island and southernmost point in Texas is nearby. Photo by Author.

and Irish influences somewhat blurred or missing. On the other end of South Texas, the Rio Grande Valley is becoming more complex as it grows in population and develops economically. According to the state's tourist map, the Valley is divided instead of being mapped as a whole functional unit. The inland parts of the Valley are in the South Texas region, but the coastal margin is included in the adjoining tourist region.

Gulf Coast

The Gulf Coast tourist region stretches along the entire coastline from Port Arthur to Brownsville. The largest place in the region is Houston, which contains an incredible array of travel and tourism infrastructure. Despite Houston's cultural attractions, the actual coastline is better represented in the literature. As a tourist region, the Gulf Coast incorporates a slew of natural features such as Aransas National Wildlife Preserve, and tourist spots such as Galveston and South Padre Island were developed. A series of notable places from Beaumont to Corpus Christi fall in this region, such as Port Lavaca and the laid-back seaside destinations of Surfside and Rockport. Tourist promotion of the region tends to ignore the urban-industrial development, and instead it focuses on ornithology, fishing, and early Texas history. One point of emphasis is the unique opportunity for bird-watching along one of North America's important migratory flyways.

Hill Country

The Hill Country tourist region includes the Balcones Escarpment and adjoining Canyonlands as well as much of the flat Edwards Plateau. Austin anchors the eastern edge of the region. The Texas-German counties northwest of San Antonio are an integral part of the region, and the correlation between German Texans and

FIGURE 11-14 Seashore, Surfside Beach
Second homes and beachcombers, Brazoria County Photo courtesy of Texas Department of Transportation

FIGURE 11-15 Dune Stabilization on Galveston Island
Efforts to vegetate and protect dunes from pedestrians, Galveston County. Photo by Author.

Hill Country is very strong. Many of the normative images for Central Texas and Heart of Texas are somewhat interchangeable with this tourism region. The geographical center of Texas near Brady falls into this tourist region. The live music scene of Austin and Central Texas, as a whole, including dance halls such as Gruene Hall are prominently represented in the tourist promotional literature. Yet, caves, bats, and Enchanted Rock also appear frequently as tourist images. However, a significant number of the images are German landscapes and events such as Oktoberfest, which provide a strong visual dimension to the literature.

254 Part Four Modern Human Geographies of Texas

FIGURE 11-16 Heart of Texas Resort
Smaller lakeside resorts on the shores of reservoirs, near Highland Haven in Burnet County. Photo by Author.

FIGURE 11-17 Tourists at Fort Worth Stockyards
International and Domestic Tourism at historic stockyards, Tarrant County. Photo by Author.

FIGURE 11-18 Alternative Entertainment in Deep Ellum
East end of Dallas has many bars and restaurants, but no residential gentrification. Photo by Author.

Prairies and Lakes

The final tourist region is the Prairies and Lakes, which stretches from Sherman on the Red River down to Shiner on the Coastal Plain. The Black, Grand, San Antonio, and Fayette Prairies support the regional name. However, no significant natural lake falls into the region, but numerous reservoirs and recreational opportunities are located there. The Metroplex is the most populated part of the region, which includes Dallas and Fort Worth, yet extends southward to include Waco and College Station. As a tourist region, it has very eclectic images ranging from bluebonnets to cattle drives. The Dallas–Fort Worth area supports a large travel and tourism infrastructure that includes numerous cultural amenities such as museums and the DFW airport. The renowned Fort Worth Stockyards is prominently represented as a real Texas experience for visitors.

In addition to tourist regions, the state has organized driving routes for touring. While not originally conceived as regions, the driving tours are meant to link historical and environmental places into a coherent theme and route. Some examples include the Independence Trail, Brazos Trail, and the Mountain Trail. Recently, the state has started to coherently designate the Texas Heritage Trails Program (THTP) and depict them as ten distinct regions. It's not clear how these ten heritage regions will interact with the seven tourist regions in the future. Moreover, the state continues to build and maintain the infrastructure for automobile tourism in the form of roads and rest areas. In an ironic twist, tourism officials have created a brochure to guide visits to historic rest areas that were constructed during the Great Depression by the Works Progress Administration; therefore, we now have tourism of tourism. These attempts to create a coherent tourist experience by designating touring routes reinforce the significance of automobiles to tourism. While the automobile is crucial to the mobility of individuals, it is also essential to the emergence of contemporary leisure. In the next section, we explore some of those playful activities associated with leisure.

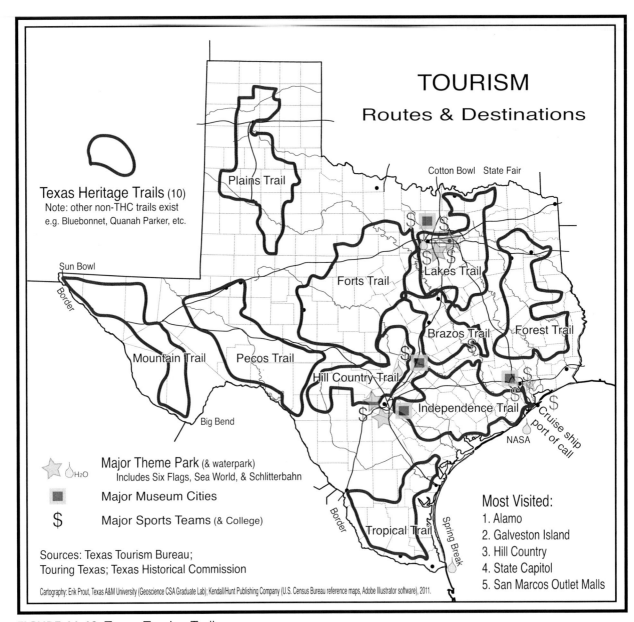

FIGURE 11-19 Texas Touring Trails

GEOGRAPHIES OF PLAY

Leisure in all its forms has numerous connections to play, and one obvious connection is the possibility of fun. Doing something, going somewhere, and being with other people are potentially fun. That's not to say one cannot enjoy doing nothing, staying put, and being alone; in fact, there is an equally valid geography of solitude. I would argue that a majority of solitude and individualism primarily occurs in the home rather than those specific public places for contemplation and prayer. Home is in many regards a safe, private realm where both fun with others and solitude of self is normal. Theoretically, geographies of the

FIGURE 11-20 Touring Signs
Both official state trails and local and regional routes are signposted around Texas. Photos by Author.

home (and the play inside them) include both a spatial pattern with other houses and the design of our social spaces in and outside of the home. For those who invest in large flat screen televisions, sound systems, and video games, the consumption or play is highly atomized in one's living room or bedroom. These individual consumptions and absorptions are typical in our society and they collectively add up to a statement of significance about American culture. However, this discussion of leisure focuses on the public and quasi-public spaces where people come together for enjoyment.

Furthermore, we must consider the terms "play" and "fun" as relative because not everyone will think in their minds and experience with their bodies my list as universal positives. Two prominent cautionary examples include football and sex. For many, Friday night, high school football in Texas is not just a game. Football is a community event with strong interpersonal relationships for those intensely involved. Like any sport, some individuals become physically injured, and there has to be psychological pressure on those who lose. The point is that while football attracts much interest and positive cheering, there are serious consequences for those who do not measure up and get left behind. Sex is an integral part of our social being and much more than a biological necessity. While being a potentially beautiful component of a consensual relationship, sexual partners do not always have identical purposes, meanings, and even memories of what transpires. The list of possible problems includes sexual abuse, sexually transmitted diseases, and relationship/breakup psychosis. In addition, conversations about sex, even academic ones, are intertwined with personal moralities and the outright inability of many to even discuss sex. Therefore, as we discuss the geographies of leisure that derive from fun seeking that are typically meant to be enjoyed, we cannot lose sight that there are downsides, consequences, and objections.

Kids' Play

The first list of playful places is titled "Kids" but it really includes people of all ages. A common theme is that most people experience these places during the daytime, and the visitors include families with members of different generations. For the young, these places are special destinations; for the old, they bring out their inner youth. In fact, the presence of different age groups supports the theoretical idea of cultural reproduction, which refers to the generational passing down of attitudes, behaviors, and beliefs. Specifically, these activities support a broader acquisition of culture because of children directly interacting with or at least imitating adults. As opposed to formal training such as a school education, these informal, fun activities impart experiences such as actual social interactions and monetary consumption with strangers. A general comment about the "Kids Play" places is the stratification of teenagers who are negotiating roles and defining spaces in between children and adults. At nearly all these places, teenagers and young adults are too old for kids' games, G-rated movies, and so on but they are not old enough for (legal) alcohol and unrestricted adult access.

Amusement Parks

The most manipulated cultural landscapes are theme parks such as Disney World. It is widely understood that Walt Disney combined the ornate gardens of Tivoli Gardens (Copenhagen) with the thrill rides of Coney Park (Long Island) when he designed the original Disneyland in California. Typically, amusement parks have multiple components including thrill rides, games of chance, fast food, and carnival strangeness. Contemporary theme parks such as Disney and Six Flags emphasize the thrills while minimizing the carnivalesque. The largest amusement parks in Texas are generally located in the large metropolitan areas: Metroplex, Houston, San Antonio, and so on. They stand apart from other amusement activities such as fairs because they occupy large permanent tracts of land and operate year around. Furthermore, they become part of the tourist infrastructure in those metropolitan areas, and at times they become linked with regional identity. Six Flags is an obvious reference to Texas' political history, but it has expanded operations beyond Texas' borders. Astroworld is another regional example, which, despite being closed, was associated closely with NASA and the city of Houston.

Shopping Malls

Another major land use that seems to be following the broad thematic design principles of theme parks is shopping malls. The mall has become the premier shopping experience in America, clearly overtaking downtowns. With the exception of major cities that have maintained exclusive shopping districts, the mall is the place to be and to shop. Initially, malls replicated the shopping experience of downtown except with ample parking and easy highway access; in fact, most of the earliest malls were straight corridors with anchor stores on each end. As the mall evolved, one design element stands out—festive. In addition to curves and multiple levels, the mall has become a sensory experience with colorful and glittery motives with banners, flags, balloons, lasers, and anything that can make the space more festive.

Rodeos, Fairs, Festivals

A host of seasonal events in Texas stand out for their connection to cultural identity. Many of these events are temporary and they coincide with a season or local historical event. County fairs are basically everywhere (in every county) and usually have a dedicated fairground that can be used for some other activity during the year. Festivals are often associated with local specialties especially around harvest time; for example, there are apricot, watermelon, and shrimp themes. Yet, festivals replicate and mimic one another toward a common American style with the same vendors operating at different events. Often in tandem

FIGURE 11-21 Modern Theater and Shopping Complex
Tinseltown in Houston is only one example of contemporary car designed rlandscapes, Harris County. Photo by Author.

with special events, rodeos are common especially in the areas of the state with much ranching. Because the actual rodeo activities are more important in many instances, the events are coordinated with the various professional rodeo circuits. The rodeo has a special resonance for many because it exhibits traits and skills associated with Texas' past and projects an image of Texas to others.

Movie Theatres

While amusement parks have large land use impacts, movie theaters are more numerous and they operate nearly everywhere in the state. The historical context for movie theaters includes the rise of Hollywood as a cultural phenomenon in America and the near universal diffusion of movies and cinema. Nearly all theaters were initially located in downtowns until drive-in theaters were invented. Those downtown theaters were set into the built environment of Main Street, and they often combined a stage to accommodate multiple uses such as drama and community events. However, the modern theater is more likely to be a multiple screen facility and actually have amusement components including video games in the lobby. In all likelihood, new theaters are located near the mall or along major transportation routes with ample parking. In a thematic fashion similar to amusement parks and malls, new theaters are designed spaces with festive motifs and playful aspects that project fun.

Playgrounds/Parks

Perhaps the most literal play place is the ubiquitous playground. They tend to be used mostly by the younger members of society; however, parents are often the initiators with the youngest as they become introduced to slides and swings. Playgrounds are found at schools, parks, and restaurants; many are designed as safe places with good visibility and single entrances so parents can mind their young with peace of mind. In the context of parks, playgrounds are only one component of the land use and landscape. The larger societal movement to create parks at various geographical scales is instrumental in the frequency and commonplaceness of playgrounds. For example, cities and even neighborhoods designate land for parks with the purpose of providing recreational opportunities. While urban parks are typically smaller, they often include basketball courts and soccer fields if space permits. Large cities often have designated and designed "green spaces" that include open areas and places to play, but increasingly they also have natural trails and urban gardens. Cities and towns often combine parks with picnic areas, and they provide tables, restrooms, and barbeque pits for groups to utilize.

Special Events—Coming of Age

Special events are in a category of their own because they signify a milestone of life. Some of the special events include Quinceañera, prom night, and even marriage parties. These special events are rituals imbedded with community tradition and status to those who are coming-of-age, and they typically happen in some public space. This particular list highlights three instances where female sexual identities transition between states. Quinceañera is the Hispanic celebration around a girl's fifteenth birthday, which traditionally signifies her availability for courtship. In Texas as well as the American Southwest, Quinceañeras have become major social events. Likewise, prom night has become a major event that capstones the high school years, and it has been portrayed as the last big social fling before adulthood. Prom night has carried a sexual connotation in America for decades as well as a more general end of innocence. Finally, marriage is a more formal change of social status that officially sanctions sexual relations in many religions and sometimes includes a name change. Marriage as a social event transpires as ceremonies and parties with their obvious consumption and symbolism to places that host these events. In addition to the wedding ceremony and reception, there are preparatory events such as bridal showers, rehearsal dinners, and even fund raisers.

Nightlife

The second list of playful places is adult oriented with a tendency towards nighttime. The common thread of these activities is the search for companionship. While sex and marriage might be the ultimate goal, most of these places are really social settings for preparation, practice, and learning about self. Friendship in the form of the known companion you arrive with is probably greater than the new potential partnership you might depart with. In many instances, these places are about intimate spaces for conversation with friends and lovers you already know. However, some of these places are designed for meeting new friends and love interests. There is a body in motion component with many of these "nightlife" places. On one end of the spectrum, people are innocently moving about creating visual stimuli and a sense of vibrancy to the setting such as a gym or a busy restaurant. At the other end of the spectrum, there are places that literally display the human body and demonstrate sexual movement such as a dancehall or strip club. Another factor of nightlife is the role of alcohol and drugs that initiate communication but can also become a way to deaden the senses and lead to other problems. In a legalistic sense, these places and consumptions are for adults despite the fact that teenagers adopt these elements in less official settings. While falling into this nightlife category, illicit activities are real and diffuse, but they are hidden to nonparticipants in ways that mask their illegality. The following list has a visible component that one can observe in the cultural landscape.

Dance Halls/Live Music

A cultural feature around Texas is the country-western dance hall with its wood dance floor and stage for live music. While the idealized form has some lively two-stepping like scenes from Urban Cowboy, the dance hall and live music scenes in Texas are varied and complex. Not all of these venues have live acts, and they are not limited to country music. Vibrant music scenes in urban and metropolitan Texas include rap and punk as well as any new trend that diffuses around the world. At the other end of the spectrum,

FIGURE 11-22 Adult Entertainment on edge of Town
Dimitris Strip Club is located in a non-residential setting along IH-45 between Texas City and Galveston, Galveston County. Photo by Author.

folk music and dancing survive in ethnic communities around Texas with polka, Tejano, and square dancing to name a few examples. Austin claims to be the live music capital, and it hosts one of the longest running public television shows, Austin City Limits. Austin exemplifies the synergy between country and rock genres with Willie Nelson and Stevie Ray Vaughn to name a few prominent performers who have graced the stages.

Bars/Coffee Shops

The difference between a dance hall that serves beer and a bar with a juke box can be very little. As a separate category, bars and coffee shops are places to hang out and spend time; as an ideal, these are places where we fit in and everybody knows our name. In reality, one can do one's own thing or one can be part of a group of people. Like other nightlife places, bars and coffee shops are third places that are distinct from both home and work. The bartender serves alcohol that can lead to mental impairment, and the barista serves caffeine that can stimulate the mind. Bars are common throughout the United States except in dry counties in the South. They differ in terms of regional décor styles, sport team affiliations, and brand preferences. For example, Shiner Bock is likely to be sold in Texas establishments amidst the banners of local collegiate teams. Because of Starbucks, coffee houses are an almost universal addition to the cultural landscapes of America, but real coffee shops as opposed to diners existed in most university towns previously as independent bean roasters and as alternative study places. Similar to their European counterparts, many of the older coffee houses serve cakes and light cuisine. Independent coffee houses are distinct because they are also sites for creative and alternative subcultures with their poetry reading nights, displays of local artwork, and a cluttered bulletin board with information about vegetarianism, homeopathic remedies, and so on.

Strip Clubs/Brothels

A component of nightlife in Texas is the aptly named adult entertainment industry. Contrary to the Bible Belt morality of the American South, Texas exhibits a Western frontier tolerance to sexually oriented businesses. Perhaps, one could state it as a healthy tension between individual liberty and community values. It could be argued that historically the American West had numerous vices such as saloons, gambling, burlesque entertainment, and even prostitution, but many communities tried to contain and control these activities. In addition, Texas is influenced by its proximity to the Mexican border. While Catholic Mexico is culturally conservative, the northern frontier/border region has zones of tolerance primarily for tourists that exempt immoral behavior—most notably prostitution. In Texas, the legacy of prostitution is still a part of our cultural memory exemplified by the movie, The Best Little Whorehouse in Texas. Along this theme, Texas has a very open strip club scene compared to many other states, often situated where single males are together for jobs away from home such as roughnecks in the oil fields, military bases, and business travelers in the cities. The range of businesses and names of strip clubs range from Gentlemen's clubs to Revues, but they are essentially an undressing of young women for money. While not necessarily having a direct connection to strip clubs, Texas also has vibrant escort services and conspicuous adult video establishments. There is no moral statement here by me; the significance is the openness and visibility of adult entertainment in the cultural landscapes of Texas.

Body Beautiful

Generally, a positive thing associated with our leisure society is the host of activities and services that make us feel better. For the most part, these things make our bodies look better, which has strong psychological benefits such as self-assurance. Critically, these things point to our obsession with appearance and consumerism and perhaps even libidinity. Many people are fixated on improving their appearance through fitness and cosmetics, but also through drugs such as steroids, surgical modification, artificial tanning, and dangerous eating disorders. The increasing number of gymnasiums, nail salons, and tanning booths are all signs of this trend. Except the gym, which actually changes one's fitness, the other places are mostly cosmetic and have a public social function. Traditionally, beauty salons have been an important third place for women because they were sites for social networks and the exchange of information. The barbershop served a similar role for men albeit with the gender differences of not calling it gossip. The emergence of day spas with complete massage and makeovers is a recent trend that warrants research because of the encompassing nature. It is already clear that we spend time and money to improve our appearances and sell our bodies.

Cruising

Essentially, doing nothing or hanging out can be a public and popular activity. To be cool, one is in the right place to be seen and to see others. While many communities have loitering laws, the place to hang out might be associated with some legal activity such as the diner, theater, and playground. A uniquely American activity is cruising, which evolved in the post WWII society of affluence with cars. Teenagers and young adults slowly drive up and down a major city street and interact with one another. Showing off one's car and acting cool are basic tenets of cruising, yet it clogged the streets with cars. It also led to other things such as racing and exhibitionism. As this form of leisure became a nuisance, anticruising ordinances were established. However, the popularity of cars and the ease of mobility will always lead to *ad hoc* encounters. It is probably fair to say that teenagers in America will always negotiate a space for themselves as they navigate from childhood to adulthood. The inclusion of automobiles into this negotiation creates new geographies of where and what teenagers do.

The impact of leisure in Texas is enormous and it is growing. The affluence and discretionary spending of many Texans will continue the consumerism that propels new cultural landscapes. The automobile will likely remain an important aspect of leisure for the foreseeable future. In Part Four, we explored the contemporary human geographies of Texas. By discussing the demographic, urban, political, economic, and social characteristics, we know where Texans live and work. In this chapter, we explored the geographical dimensions to leisure and hopefully developed an appreciation of tourism and the role of play in consumption. In Part Five and Chapter Twelve, we return to the concept of region that underpins our discussion of Texas. Finally, we explore the macro-regions of Texas by synthesizing the cultural and physical geographies.

Part Five
Regional Geographies of Texas

PRELUDE TO A CONCLUSION

Where are we? Who are we? What are our places like? How should we describe our identity, territory, and nature? So far, our geographical inquiry has asked these questions about Texas. We began our inquiry with some initial definitions and parameters of both Texas and geography. Our journey took us through the cultural diversity and historical evolution of Texas. Then, we traveled over the literal Earth as we discussed the physical geography of Texas. We continued by exploring the contemporary human geography of the state. Now, we elaborate our discussion by returning to the concept of region and (re)constructing the major regions of Texas. In addition, we compare and contrast academic and popular regions. Finally, we conclude with a brief foray into the future and speculate about the trends.

Texas is big—both in reality and in our minds. The large land area of Texas is a constant component of its geography. The long distances between places are part of the explanation of Texas' cultural and physical diversity. Texas is also big in our imaginations. First, there was Empire, which flourished during the Republic, and movies such as Giant have propelled the mythology of Texas. While the meta-narratives about Texas tend to be unifying and grand, Texas is in reality a composition of fragmented narratives. The distinction often depends on whether one sees and emphasizes the whole blanket or the quilting together of different stories that compose it. In addition, the narratives of Texas are embedded in another "one-or-many" meta-narrative: the United States of America. Texas is fundamentally a part of America and the story of America, and we can unmistakably see America in the places and landscapes of Texas. We can also see traces of the rest of the world—especially Mexico—in the peoples and places here in Texas.

PLACES AND REGIONS

Place is one of the more important concepts in geography. For geographers as a whole, place generally corresponds to scale—a very manageable scale that transcends physical and human geography. In this view, place is a setting or site; it roughly corresponds to what we can see around us. For the soil scientist, place is the literal study site where she digs a profile as well as takes into account the environmental context of terrain and vegetation. For the landscape photographer, place is the visual viewscape that can be captured in a single snapshot or a collage of images on a specific topic. In addition, there is general understanding that place and region are connected, and place occupies a smaller area than region. Some scholars would impose a more rigid interpretation that places are embedded in regions. For example, a place such as Llano is in the Central Texas region. While this continuity of scale works toward a common understanding between geographers, there is a significant subgroup that treats place as a more fundamental idea. In particular, cultural geographers conceptualize place as having meaning, and these meanings are insights into our individual existence and collective humanity. Perhaps the most meaningful connection is the relationship between culture and place. Not only are places and cultures understood through each other but they also literally form and evolve together. Another meaningful connection is the relationship between space and place. In this perspective, place is qualitatively different and contains the meanings from being known and experienced.

Sense of place is a term that cultural geographers use to describe one's relationship to place. It is often intended to mean the personal experience of place and even the feel and feelings one has in a place, which implies the literal use of the senses and environmental perception. For example, standing alone in a rainforest one might recall the cool, damp, quiet, and outright loneliness of the experience; a contrary sense of place could be a hustling-bustling urban area with exotic smells and sounds. What about extremely typical places like a McDonalds or a Walmart? Some scholars and social critics call these **placeless** because they exemplify homogeneity of landscapes and even experiences. In this perspective, only unique places have a sense of place. Inauthentic and overly standardized commercial sites are placeless, and placelessness is the way to interpret their landscapes. I personally argue that all places have a sense of place, and therefore a place can have a positive or a negative sense. Therefore, placelessness is simply a negative sense of place. One last point, much of the descriptive writing on places is reflexive because the authors relate their journeys and encounters while being in a particular place. This form of place writing is common with travel writers and promotional tourism brochures. Academic writing about place can also have a reflexive component. However, academics try to put their experiences into the appropriate context and significantly with other texts. Expectedly, sense of place is almost always a qualitative type of research—closely related to ethnography and landscape interpretation. Examples in Chapter Twelve include a few box descriptions on specific places such as Luckenbach that follow this reflective description.

LEARNING OBJECTIVES: CONCLUSION TO THE GEOGRAPHY OF TEXAS

Part Five of this textbook is associated with the big ideas of region and explores the relationship of Texas to regions at different geographical scales.

- Discuss the major regions of Texas

    You should be able to identify the major regions in Texas and know their locations and significant places in them.

    You should be able to summarize the cultural and physical generalizations for each of the major regions.

    You should be able to elaborate and expand on the major region your hometown falls into in terms of how representative it is as a place in the region.

    You should be able to elaborate on the ideas concerning regions.

    You should be able to discuss Texas as a part of the United States, North America, and the World.

- Maps: Familiarity with various regional maps of Texas (Jordan, Meinig, Prout)

- Spatial Data: Population size of major places in major regions.

- Definitions:

    Place
    Sense of Place
    Perceptual map
    Perceptual region

"University as Place"

FIGURE 12-1 Aggieland, USA
Texas A&M campus, College Station, Brazos County. Photo by Author.

12
Regions of Texas

Texas is its own region, yet there are numerous other regional associations one considers with Texas. On a more global geographic scale, Texas is part of the United States and North America. It is interconnected and interdependent with the entire world in terms of trade, communication, and environmental pollution. Significantly, it is in the American Southwest along the Mexican border, which shapes and directs some of our interconnections. In addition, its Anglo roots lie in the American South (both upper and lower), and its frontier attitude corresponds with the American West. Interestingly, things like cattle ranches, oil pump-jacks, and Friday night football both connect and distinguish Texas from our fellow citizens. These are things found and celebrated in other parts of the United States, but they somehow become emblematic of Texas. Applying Texas to a single American subregion isn't so simple because it is not typical Southern, Western, or Southwestern. It's Texas.

At a more local geographic scale, Texas is complex with many heterogeneous elements. Texas has ecological diversity, cultural pluralism, and its large size. Depending on the element, geographers can discern distinct regions within Texas. There are cultural geographies based on language, religion, and settlement. There are physical geographies based on geologic, climatic, biotic, and hydro realities. There are human geographies based on demographics, urbanization, politics, economics, and even leisure. In this final chapter, we return to the big ideas of regional geography and discuss the regions of Texas. First, we introduce perceptual regions that ordinary Texans regularly use to describe their home regions. By examining popular knowledge of regions, we can grasp the geographical diversity of the state in yet one additional way. Sometimes the difference between academic and popular generalizations can be quite large. Second, we try to synthesize a description for the major regions of Texas. Perhaps, the finest art of regional geography is to describe large, complex regions. As part of these large regions, selected places are described in more detail. It is in places that we actually experience the world. Moreover, much of our personal knowledge of a region comes from the specific places we live, work, and play in. Finally, we conclude with some possible interpretations of future regions for the state. Future Texas regions will reflect both continuities with the past and the present trends of change.

POPULAR REGIONS

Local inhabitants have many different names for where they live. They can draw on more than official names and use neighborhoods, roads, physical features, and even ethno-racial characteristics. At the local scale, popular regions can become extremely complex as individuals may have their own set of names and style of usage. As the geographic scale widens beyond a single county, popular regions become more observable to an outsider. Typically, a single name or characteristic for the region becomes popularly known, accepted, and even displayed on road signs in the landscape. Three dimensions to popular regions are perceptual, vernacular, and historical. Perceptual dimension is the mental map idea, whereas every

FIGURE 12-2 Last Picture Show Revival
IMAX Theater downtown San Antonio near the Alamo with high tourist and shopping presence, Bexar County. Photo by Author.

individual has one's own geography in their head. The vernacular dimension includes those more localized names that inhabitants use. The historical dimension refers to the resilience of previous regional notions by earlier generations, especially environmental associations with the land.

Perceptual regions are regions that are perceived by ordinary people. For our purposes, perceptual regions include both lived-in regions by local inhabitants as well as regional subdivisions of the state by all residents. For a large state such as Texas, it is normal to perceive logical parts that happen to coincide with cardinal directions. *Vernacular regions* are regions that local inhabitants believe exist; however, they usually do not appear on maps. Recall that some of our vernacular examples include Dixie and Aggieland. In addition to Dixie, we could consider other Southern labels such as Cotton Belt and Bible Belt applicable to Texas. Regions with popular labels such as Alamo and Capital are similar to vernacular because they do not reflect the actual place-names. The third dimension to popular regions is historical usage, which may date back to the early Hispanic and Anglo settlement. Many regional names in Texas with a historical origin do not necessarily have the same resonance today because they were based on past human–environmental relations. Nevertheless, historical usage often survives without a compelling reason to change.

Perceptual Regions

The study of perceptual regions never advanced much despite *ad hoc* attempts to create research projects. Terry Jordan's work on Texas is one of the few examples in the literature. We know people have mental maps, but trying to scientifically collect enough perceptions to signify collective trends is very tedious. Jordan always qualified his results because of the obvious methodological difficulties and sample set biases. Yet, Jordan had the advantage of having conducted enormous fieldwork in the state, and he could deduce that he was receiving accurate results. The perceptual geography survey was primarily conducted at universities using almost 4000 students. Participants were asked a series of basic questions to ascertain their hometown location and the names they used for their own region.

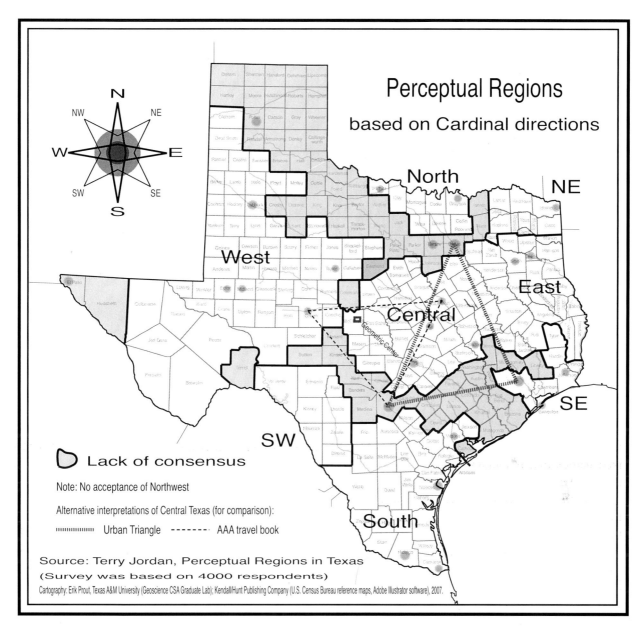

FIGURE 12-3 Cardinal Direction Regions

The results reveal approximately 37 perceptual regions for Texas. Yet, they need to be divided into two categories. The first eight perceptual regions are associated with cardinal directions. Considering the size and shape of the state, we would expect some directional terms that correspond with the cartographic images of Texas. The four main cardinal directions, North, South, East, and West are all perceptual regions. For the most part, they align with the geographical reality except North Texas, which is centered between the Metroplex and the Red River. While not technically a direction, Central Texas exists as a perceptual region in this framework. In addition, three of the four combinations of cardinal directions appear as regions. Northeast and Southeast are literally the respective northernmost and southernmost counties of East Texas. Southwest Texas appears as a perceptual region in a few counties southwest of San Antonio. No significant use of Northwest Texas by respondents occurs in the survey; for many, the northwest is missing in the shape of the state. While not appearing in this survey, by oversight or response, North-Central, East-Central, and Far West Texas are used in certain contexts.

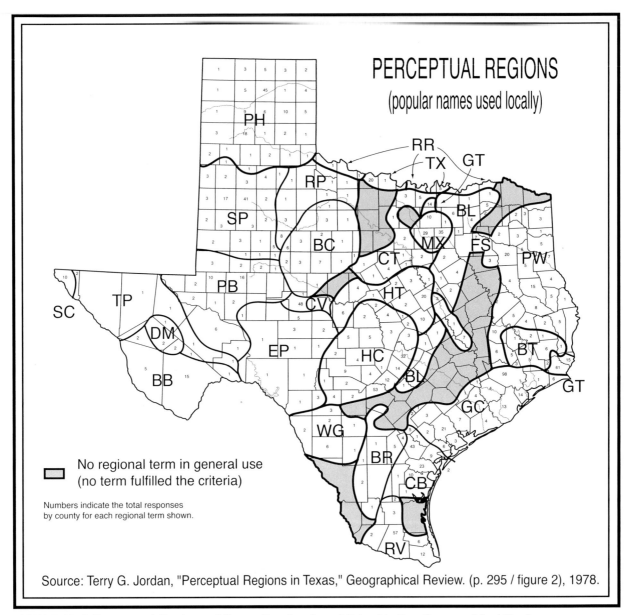

FIGURE 12-4 Perceptual Regions of Texas Previously published in the *Geographical Review*

The second category of perceptual regions is the more complex pattern with its vernacular and populist names. In total, there are 29 of these perceptual regions. All of the regions can be classified into categories, but some have multiple connotations. The primary classification is listed in Table 12-1 as type of name. Nineteen or a majority of the regions have environmental names. Many of these names were derived from early Anglo-Southern migrants as they sought to create agricultural settlements. The names reflect the environment in a variety of ways including terrain, water, flora, soil, and relative location. Ironically, a few of them are not consistent with the actual surface geography. For example, the Permian Basin is named for the geological structure underground that provides oil and not the flat surface plain. For that matter, none of the valleys are technically valleys; they are really floodplains. References to soil color and vegetation are less relevant today. While the name Blacklands survives, Redlands in East Texas has nearly disappeared as a common response. Coherent stands of oaks in parts of the Cross Timbers are difficult to find, brush removal for ranching is common in Brush Country, and the dense vegetation of a

TABLE 12-1 Perceptual Region Names

Vernacular region	Type of name
Big Bend	Environmental (and promotional)
Big Country	Promotional
Big Thicket	Environmental (flora)
Blacklands	Environmental (edaphic)
Brazos Valley	Environmental (river)
Brush Country	Environmental (flora)
Coastal Bend	Environmental (Gulf)
Concho Valley	Environmental (river)
Cross Timbers	Environmental (flora)
Davis Mountains	Environmental (terrain)
Edwards Plateau	Environmental (terrain)
Free State	Political-historical
Golden Triangle (SE)	Promotional
Golden Triangle (N)	Promotional
Gulf Coast	Environmental (Gulf)
Heart of Texas	Promotional
Hill Country	Environmental (terrain)
Metroplex	Political (census classification)
Panhandle	Political (geometric)
Permian Basin	Environmental (geology)
Piney Woods	Environmental (flora)
Rolling Plains	Environmental (terrain)
Red River	Environmental (river)
South Plains	Environmental (terrain)
Sun Country	Promotional (good climate)
Texoma	Political (and promotional)
Trans-Pecos	Environmental (relative location)
Valley (Lower Rio Grande)	Environmental (river)
Winter Garden	Promotional (good climate)

Source: Terry G. Jordan, "Perceptual Regions in Texas," Geographical Review, 1978.

thicket are confined to small parts of the Big Thicket. Regardless, the resilience of environmental names is impressive.

Four perceptual regions are classified as political. Panhandle is a cartographic legacy resulting from the state's political boundary shape with the Compromise of 1850. Of course, Panhandle can come to connote the environment of the High Plains as well. Texoma (Texhoma) is a concocted hybrid name along the Red River political border between Texas and Oklahoma. With the large reservoir and recreational infrastructure, Texoma is a regional name also associated with promotional activities. Free State is a vernacular found in Van Zandt County that dates back to the Civil War. However, the Free State slogan appears elsewhere, usually as a libertarian statement such as in McMullen County. The final political region is the Metroplex, which we also know as a large functional region. The term Metroplex derives

FIGURE 12-5 Heart of Texas Billboard, Brady
Use of a vernacular name in advertising, McCulloch County. Photo by Author.

FIGURE 12-6 Free State Sign, McMullen County
Libertarian sentiments in hand-made road signs. Photo by Author.

from the 1970 Census when Dallas and Fort Worth were first classified as a Consolidated Metropolitan Area. The term morphed into a single word and eventually became popular enough to be used as a promotional term for the entire urban area.

Six perceptual regions are classified as promotional, which seems to be a growing trend. All six are invented or concocted to attract economic investment or retirees. Two separate places have initiated the term Golden Triangle. One is located in Southeast Texas along the Upper Coast; this triangle is anchored by the cities of Beaumont, Port Arthur, and Orange. The other Golden Triangle is centered on Denton, where the two IH-35s junction North of Dallas and Fort Worth. A part of the Coastal Bend has started to call itself the Golden Crescent; despite using the name for its Council of Government, it hasn't gained popular acceptance to appear in the survey. Sun Country in El Paso has somewhat become popular enough, and Winter Garden southwest of San Antonio including Del Rio and Eagle Pass does as well. Both are trying to reinforce their climatic appeal to senior citizens looking for warmer and drier climes to spend their winters away from the northern latitudes of North America. Of course, the (Rio Grande) Valley also has that popular connotation without creating a fictional name.

Big Country and Heart of Texas with their fictional names are the last perceptual regions classified as promotional. Centered on Abilene, Big Country occupies a significant part of West Texas. Interestingly, the Heart of Texas perceptual region and Council of Government boundaries are slightly different. While these two regions are promotional, many of the other regions have a promotional dimension. One aspect of promotion is name recognition, so Texoma and Concho Valley are trying to compete for everyone's attention in the crowded marketplace of leisure and development. Places such as Big Bend and Big Thicket are crossovers into promotional because they are natural attractions that attract outsider's attention. Even perceptual regions such as Hill Country and Gulf Coast are utilized as tourism regions. The role of Councils of Governments cannot be overlooked in the evolution of perceptual regions. As these regional governments try to promote economic development, they create names for promotional purposes. The use of golden, heart, and big are positive spins on their respective regions, which are likely to become accepted by local residents over time. Even a name like Deep East Texas, which implies a "Southern" attachment, has gained usage at the expense of Southeast Texas and even Big Thicket. One wonders if this survey was conducted today, what responses would appear, disappear, and migrate?

Academic Synthesis

The scholarly perspective on popular regions is a circumscribed one. In theory, an informed scholar may help governments make a promotional name. For example, individuals at Texas A&M University have helped promote the term Research Valley. For the most part, there is no popular acceptance or usage of the term in Brazos County or any evidence that it is displacing Brazos Valley or Aggieland as a popular regional name. Clearly, scholars cannot dictate popular regions or any toponym for that matter, but scholars can research the popular use of vernacular and idiosyncratic names. The previously mentioned survey of perceptual regions is one such attempt. Another scholarly approach is to synthesize all the variables and to delineate on a map the broad generalization of regions.

Donald Meinig's work not only provides a contemporary framework for the major Texas regions, but it gives one insight into the evolution of the major regions. In 1860, there were four broad Texas regions: Central, East, South, and North. Significantly, Central Texas as a concept was initially located on the Coastal Plain, and it included the gateway of Galveston–Houston. East Texas was the area adjacent to Louisiana and Arkansas. South Texas was the area of strong Spanish and Mexican influence. North Texas was the area North of Central Texas extending to the Red River border. This historical notion of major regions explains why North Texas today is still associated with Dallas and Fort Worth instead of the Panhandle. In 1860, there was no perception of a West Texas, and West Texas region comes about only after settlement of the Great Plains frontier.

A hundred years later in 1960, Meinig delineates nine major regions for Texas. Three of the four initial regions are still in the same relative location. Central Texas has moved slightly inland toward the geometric center of the state. Along the coast, he argues for a Gulf Coast region with its urban and industrial developments. West Texas has become a major perceptual region and actually occupies more area than other regions of the state. In addition, Meinig introduces the Panhandle, German Hill Country, and Southwest Texas as major regions. However, his Southwest Texas region does not coincide with the perceptual region with the same name in the state; Meinig's Southwest Texas better fits into a broader U.S.–Mexico borderland conception.

Terry Jordan also provides a synthesis for traditional rural culture areas in Texas. As a scholar who documented the migration and diffusion behind Texas' cultural diversity, Jordan interpreted the cultural landscape for rural Texas in a very comprehensive manner. With this map, he elaborated on the slight variations in the major culture groups. Therefore, we can analyze the map for any underlying patterns that might explain the contemporary popular regions. Similar to Meinig, Jordan includes a region for the German majority in parts of Hill Country but a smaller Panhandle region dominated by Lower Midwesterners.

The adjacent Upper Southerners region is divided into three more nuanced zones. The easternmost part of the Upper South realm was occupied earliest by middle-class farmers attracted to the dark soiled

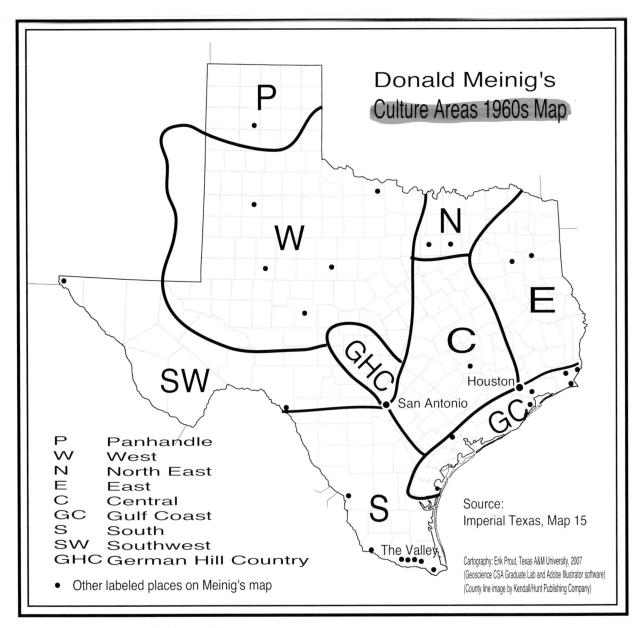

FIGURE 12-7 Meinig's Regions of Texas

Prairies. The adjacent area was settled by poor whites who had cultural continuity with Appalachia. The westernmost part was settled primarily by later generations of Upper Southerners from the previously occupied realms in Texas. The Lower Southerners region is also divided into three more nuanced zones. A plantation aristocracy with their slaves settled the large eastern area. Along the Gulf Coast, the original plantation aristocracy from Louisiana settled, but it became less distinct because of additional migration from many different points of origin by various culture groups such as Lower Southerners and Europeans. Later, a small pocket of poor whites eventually settled in the Big Thicket region,

The area Jordan called a Shatterbelt (in his ethno-cultural regions map) also can be discerned into two regions. German-dominated parts of Hill Country are differentiated from the more heterogeneous Mixed region. The Mixed region with its ethno-cultural pluralism strongly correlates with Meinig's contemporary Central Texas. The region displayed much bilingualism and cultural diversity as Mexican, European, and Southerners settled in a patchwork pattern. Lastly, Jordan depicts a continuous belt of

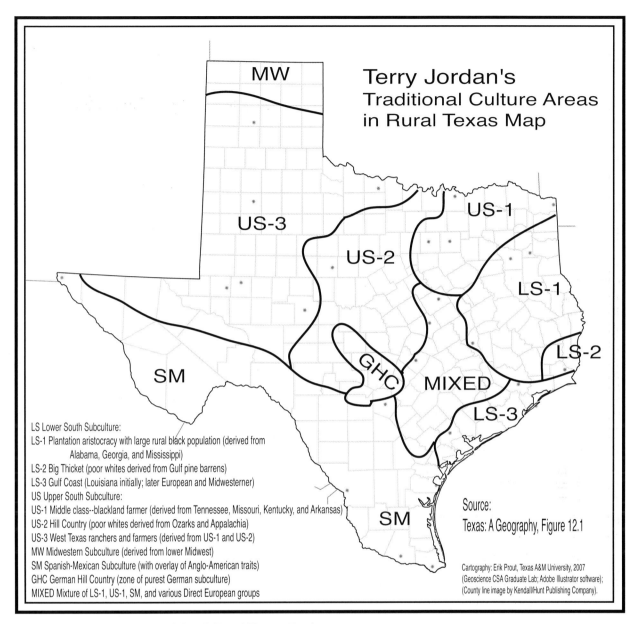

FIGURE 12-8 Jordan's Traditional Rural Texas Regions

Spanish/Mexican culture along the current U.S. border. Jordan's map generalizes the state into ten regions with common migration origins, which initially had strong continuities of their respective cultural landscapes. The early settlers in each region had similar ways of life, folk architectural backgrounds, political identities, and environmental perceptions that influenced the decisions they made about how to settle. Despite modern changes, these rural landscapes persevered to some degree.

MAJOR REGIONS OF TEXAS

There are numerous ways to conceptualize the regions of Texas. For a state as large and populous as Texas, thirty to forty popular regions are reasonable. Moreover, popular regions are an empirical truth because they carry the ethnographic reality of the people who know their own regions and use those

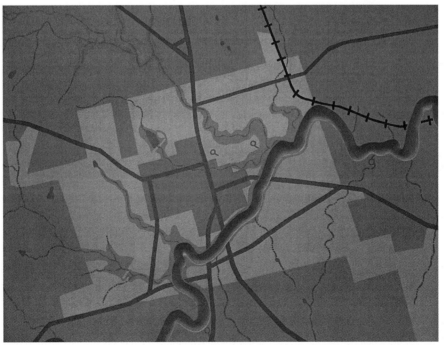

FIGURE 12-9 Writing on the Wall: map in alley
An art project that depicts the local history with outdoor painted wall maps: in the alley across from the Lampasas County Courthouse. Photos by Author.

names. Nevertheless, it is extremely fragmented, and too many regions are difficult to grasp for many students and scholars alike. A large number of regions are also difficult to draw in terms of cartography and maps. The task of generalizing and simplifying all this diversity and complexity is possible with the concept of region. Synthesizing all our knowledge about the state, the final map that I propose has only ten major regions (Figure 12-8). Other generalizations are equally valid ways to conceptualize Texas depending on the audience. For example, the tourist regions' map has an even simpler scheme but with promotional names and contestable borders (Figure 11-8). Jordan's map based on migration is extremely useful, but it does not incorporate urban areas (Figure 12-8). In the end, the *Major Regions of Texas* map I propose is visually more similar to Meinig's map (Figure 12-6) than any other. In the next paragraph, I clarify the conventions used to create this map.

Five of the ten regions utilize the cardinal directions for their names because they are both inherently understood as a framework and popularly used by local residents. In addition to being known regions, I use the five regions as my outline for description. The five regions are Central Texas, South Texas, East Texas, North Texas, and West Texas. On this map, these five regions are drawn as zero-sum areas with no place being in more than one of these five regions. Two small areas do not fit very well into this framework: along the Gulf of Mexico, which was historically part of Central Texas, and the northernmost Panhandle. Four additional regions, German Hill Country, Panhandle, Gulf Coast, and Borderlands, are drawn as areas that may overlap those five other regions. Therefore, it is possible for some places in Texas to fall into two or three or even four major regions in this scheme. The tenth and final region is Urban Texas, which is portrayed cartographically as non-contiguous nodes that obviously fall into other regions. For simplicity, only the metropolitan areas surrounding the largest urban areas have this region's symbolism on Figure 12-8. A select few other medium regions and large places are identified and labeled on the map for reference and as part of the major region's discussion.

Central Texas

Central Texas is a much more complex region than simply situating itself between the East-West and North-South continuums. Central as a term carries connotations of middle, core, and centrality, which implies in this context Texanity. Can a region be most Texan? I would cautiously answer "no", yet it is far away from any of the cross-border similarities that we find elsewhere. An analysis of how Central Texas represents itself shows strong normative images, but are they accurate stereotypes? In particular, the Heart of Texas popular region represents itself as rodeo-loving, flag-waving Texans. Physically, Central Texas straddles the humid-arid climatic divide, and as a region has to deal with both flooding and drought hazards. Central Texas generally has mild mid-latitude winters that typically lack long periods of hard freeze. At a much localized scale, proximity to rivers with their immediate floodplains and riparian ecosystems is hugely important. The flash-flooding scenario in the canyons along the Balcones Escarpment exemplifies this as well as the historic cotton farming in the river bottoms of the Coastal Plain.

Historically, Central Texas included the coastal gateway cities of Galveston and Indianola, and we could think of the Austin Colony (Empresario) and the Mexican Department of Brazos as precursors. In a way, Central Texas was oppositional to the more Southern U.S. characteristics of East Texas and the more Northern-Mexican characteristics of South Texas. In fact, one of the features that academics emphasize about Central Texas is the high degree of cultural diversity because of the mixing of different migrant groups on the middle Coastal Plain. Contemporary popular understanding of Central Texas pulls the region toward the geometric center of the state (Figure 12-2). In addition to Heart of Texas, numerous perceptual regions such as the Brazos Valley and parts of the Blackland Prairies and Edwards Plateau correlate with Central Texas. If we exclude San Antonio and Houston along the edges, Austin is the largest place in Central Texas; then, Killeen/Temple, Bryan/College Station, and Waco round out the metropolitan areas.

280 Part Five Regional Geographies of Texas

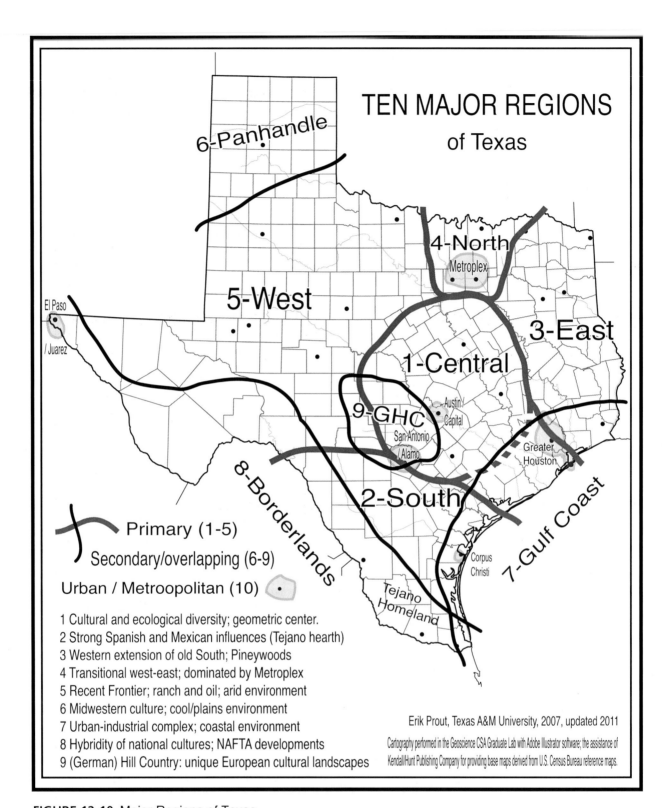

FIGURE 12-10 Major Regions of Texas

EVERYBODY IS SOMEBODY IN LUCKENBACH TEXAS

Luckenbach might be one of the best known little places in Texas. In reality it is so small with only 25 residents; it probably wouldn't register as an official town except for its notoriety. Luckenbach's most famous resident is Willie Nelson, who has included Luckenbach in his lyrics as well as implying it's a special place for music-making (see Waylan Jennings Luckenbach Texas: back to the basics of love). However, the place-name toponym reveals its origins, with German-speaking settlers. The generic name bach is the German word for a small stream or creek, and the specific name Lucken is a proper name of an early family (not luck as in good fortune). Located in Gillespie County approximately ten miles from Fredericksburg, Luckenbach consists of an old post office that simultaneously served as a country general store. Today, it is a tourist market with a bar in back; occasionally, a musician is playing a guitar and singing a song. Outside the building is an outdoor picnic area with obvious BBQ and live music arrangements, and on the other side, an oversized parking area. It's obvious that Luckenbach can accommodate many more visitors than residents.

FIGURE 12-11 Old Post Office in Luckenbach
Small town where everybody is somebody, Gillespie County. Photo by Author.

German Hill Country

Much of the vernacular Hill Country is becoming synonymous with Central Texas, and the popular usage seems to be expanding beyond its origins along the Balcones Escarpment to encompass more and more of the Edwards Plateau. Academics consistently recognize the European contribution to Texas, but most European groups were outnumbered by Anglo-Southerners. The regional exception is the area Northwest of San Antonio along the Balcones Escarpment and stretching inland. In fact, a dozen counties had a German-speaking majority for many decades after settlement, and they retained cultural traits from Europe the best. Some of the large Lutheran concentrations as well as the Freethinker communities were found here. The expression German Ten refers to the political geography of thinking about German-Texans as a voting constituency. Because of their numerical dominance in Hill Country, geographers see a strongly German-influenced region defined by the unique folk cultural landscapes. German Hill Country is drawn to include the upper Guadalupe and Medina Rivers, which includes New Braunfels, Fredericksburg, Kerrville, and Bandera.

Gulf Coast

The area along the Gulf of Mexico, especially the lower Brazos and Colorado Rivers, was initially part of Central Texas. Meinig describes the contemporary Gulf Coast for its urban-industrial character that derived from oil and petrochemical activities. Houston is the prototypical place on the Gulf Coast because of the transformation of the "Bayou City" into a major metropolitan area. The inclusion of the entire Upper Coast is appropriate in this regard. However, the applicability of the southward extension is less so as it eventually tapers out as urbanization and industrialization become less dominant and even scarce in the landscape. The other component to consider is the human–environment relationship along the Gulf of Mexico. The Gulf waters support a viable fishing industry along the entire coastline; Texas has numerous fishing villages. More recently, the numerous bays, beaches, and islands have developed into recreational regions with wildlife refuges, vacation homes, and the attendant development. This environmental notion of the Gulf Coast is stronger in the landscapes of the Coastal Bend and South Texas. Therefore, I've drawn the Gulf Coast region to encompass the entire coastline, yet with the understanding that there is a transition between Port Arthur and South Padre Island. In addition, the Gulf Coast is drawn to overlap the adjoining edges of both East Texas and South Texas. The primary association for the area between Houston and Port Lavaca must be Gulf Coast because the historical notion of Central Texas hasn't survived.

South Texas

South Texas is commonly associated with its Spanish and Mexican roots. In many ways, South Texas is the first Texas. The successful settlement of San Antonio and La Bahia began the process of large-scale European diffusion to Texas. Moreover, this settlement on the Coastal Plain was actually called Tejas, which evolves into the contemporary name Texas. San Antonio was the entropôt of trade and communication with Mexico, and it has traditionally been considered the dominant city of South Texas. Yet, San Antonio was also a launching point for settlement of Hill Country and the western frontier. As a major city, its contemporary urban qualities and geographic scale require us to conceptualize it as part of the Urban Triangle and not separate it from the growth along IH-35 corridor up to Austin. The state has recently responded by separating the economic data of San Antonio from South Texas, which prevents the San Antonio data from skewing the overall region.

South Texas is drawn to include the San Antonio River Valley, which places Gonzales and Victoria on the edge of the region with Central Texas. Corpus Christi is both in South Texas and part of the Gulf Coast region. This drawing of South Texas includes the perceptual region Coastal Bend. In the other direction, South Texas is drawn along the Balcones Escarpment to Del Rio and the Mexican border. This includes the handful of counties that identify with Southwest Texas and the promotional name Winter Garden. The remaining bulk of South Texas is the lower latitudes in Texas that incorporates the perceptual regions Brush Country and the Lower Valley.

FIGURE 12-12 Spring Break Landscapes of South Padre
Main Street South Padre Island, Cameron County. Photo by Author.

SPI is known for its Spring Break festivities, so I expected to see my own students walking around. Maybe because it was early summer, it seemed rather empty. It was obvious that many of the condos were vacant, and quite a few businesses were closed. The barrier island environment was completely hidden except for the boardwalk bridges over the vegetated sand dunes. Along the beach, numerous life guard stands and warning signs were present. It was easy to see the party infrastructure that has made this place famous.

Borderlands

The U.S.–Mexico border is unique because of the socioeconomic contrast combined with intense interaction found there. The "Border" is more than a 2000 mile–long line on a map; it is a region in its own right because of its different characteristics from both the United States and Mexico. Interestingly, a series of twin cities pairing up along the entire border is one empirical characteristic. In addition, there are linguistic differences such as high rates of bilingualism and the presence of pidgins such as *Spanglish*. One cannot escape the reality of multiple controls on the movement of people with the large presence of border police and identity checkpoints. At another level, the border is a liminal or almost surreal space with the harsh clash between imaginations, expectations, and realities. The expression of liminality occurs in different places and settings, yet seems to be centered on individuals experimenting with being or doing something different. For Americans, the border is where they experience the Third World other, and they can see stereotyped Mexican landscapes and dabble in their personal taboos. For Mexicans, the border is where they experience some of the opportunities of America and potentially break away from their cultural norms.

SUBLIME LAREDO

Laredo, Texas, sits across the Rio Grande from Nuevo Laredo, Tamaulipas; however, both cities think of themselves as really one. Like other twin cities along the border, Laredo is a very complex functional region with specialized and different activities on each side of the political border. As the premier gateway between the U.S. and Mexico, Laredo has expanded enormously on both sides of the border because of the economic changes associated with NAFTA and, more broadly, globalization. Amidst all the hustle and bustle of the border, some very distinct places are discernible. Not many of these places are quiet and quaint; they are defined by their uniqueness from everyday America and Mexico. Downtown Laredo close to the border is a series of shops and services oriented to the cross-border traffic of Mexican workers and timid American bargain hunters. Away from the old downtown, Laredo sprawls out along major transportation routes with trucking and storage facilities. In Nuevo Laredo, the city center includes both the CBD functions for its residents as well as those consumption services for American tourists. Along the periphery of town, Maquiladoras appear like a series of business parks in the United States would. The inner Laredo–Nuevo Laredo urban core reflects the cultural continuity and services for frequent border crossers, but the outer Laredo–Nuevo Laredo periphery reflects the heavy lifting of the global economy.

FIGURE 12-13 Streets of Laredo along Border
Nuevo Laredo, Tamaulipas and Laredo, Webb County are one metro area; pedestrian oriented businesses are abound near the older border crossing bridges. Photo by Author.

The Borderlands region stretches the entire length of the border overlapping the margins of both South Texas and West Texas. In South Texas, the rapid urbanization of the Lower Rio Grande Valley is very apparent and is creating a coherent and contiguous metropolitan area between Rio Grande City and South Padre Island. Academic contemplation of South Texas has produced the idea of a *Tejano Homeland*, which argues that it is more than a hybrid zone of Mexican and American cultures. The Borderlands and Valley typify the homeland concept because it serves as a permanent base for Mexican-Americans or Tejanos in Texas, and it supports the continual Mexican migration to the United States.

East Texas

East Texas is typically the strongest perceptual region of any major region. In surveys, East Texas is the most correctly identified major region for non-residents; it may help that East Texas is actually the easternmost part of the state. There seems to a strong consensus that East Texas exists, which probably reflects the more universal awareness of its different characteristics. Historically, East Texas was the periphery of Spanish control, and it was the most Anglo of the Mexican Departments (Nacogdoches). The association with the vernacular name Piney Woods is very strong, and the pine forests of East Texas are one of the dominant landscape characteristics. For academics, East Texas is the western extension of the Lowland South. For example, the coniferous forest of the Southeast stretches westward over the Mississippi River and finally transitions to oak forest in Texas. Moreover, Lower Southerners migrated

FIGURE 12-14 Quaint Preservation in Jefferson
Old towns in East Texas have plenty of Southern charm, Marion County. Photo by Author.

Known for its Bed and Breakfast establishments, Jefferson attracts a curious visitor type. And for the most part, that Southern charm was on display in both the historical courthouse area and the residential neighborhoods. Walking into a sports bar made me think I was in Louisiana. The TV stations were out of Shreveport; there was Mardi Gras and LSU Tiger posters. Was this an example of cross-border similarities?

to the river valleys of East Texas in an attempt to transplant their plantation economy. Meinig once described East Texas as a relatively stable black/white society. Like other parts of the old South, unfortunately, East Texas has had its share of tense racial politics and incidents.

East Texas is drawn along the Trinity River between the Metroplex and Galveston Bay, which was the traditional boundary ascribed to East Texas. An alternative approach is to use IH-45 between Dallas and Houston. The region continues northward to the Red River with the line drawn through Paris. The entire Texas border with Louisiana and Arkansas is part of East Texas, and, historically, those states were important passage routes for Southerners migrating to Texas. The popular acceptance of East Texas and Piney Woods labels does not exclude other names. There is a popular usage of both Northeast and Southeast Texas. It would be too easy to conceptualize Northeast and Southeast as just the northernmost and southernmost parts, respectively, of East Texas. While Northeast Texas is not typically used as a promotional name, the tristate area currently promotes itself with ArkLaTex, for example. Texarkana is a medium-sized city that falls into both Texas and Arkansas, and it has a unique postal code, Texarkana USA. Southeast Texas does promote itself despite a complicated picture with vernacular use of Big Thicket, Deep East Texas, and Golden Triangle. The term Southeast is based on Beaumont and the adjoining areas working together to promote recreational development. The Southeast in terms of Beaumont, Port Arthur, and Orange is also part of the urban-industrial Upper Coast with its ports and petrochemical plants. Therefore, Port Arthur can be considered Southeast, Upper Coast, Golden Triangle, and included in both the East Texas and Gulf Coast regions on Figure 12-10.

North Texas

North Texas is the smallest of the cardinal direction regions, but it has persevered through the changing borders of Texas. Historically, North Texas was settled by Upper Southerners looking for good farm land. Some settled along the Red River, on the Blackland Prairies, and eventually into the Cross Timbers. Separate from the Lower Southern East Texas and North of the Empresarios of Central Texas, the region developed its own identity. Ironically, Dallas grew into a large city because it played a dominate role in the Northeast Texas cotton trade and East Texas oil fields. In contrast to Dallas, Fort Worth spread its influence westward, capturing a large share of the cattle marketing and meat processing activities of West Texas. Academics tend to conceptualize North Texas as a transitional zone between East and West Texas, and it lacks an obviously clear boundary with Central Texas. They also like to point out that North Texas is not the northernmost part of the state. North Texas is drawn to include the Dallas–Fort Worth Metropolitan Area and extending northward to the Red River. The popular regions that overlap with North Texas include Blacklands, Cross Timbers, Red River, Texoma, Golden Triangle, and the Metroplex.

Urban Texas

The Metroplex epitomizes the contemporary Urban Texas region because it stands out for its population size. As the largest metropolitan area in Texas, around one-fourth of the state's population lives in this single Metropolitan Statistical Area centered on Dallas and Tarrant Counties. The rise of urban identity subsumes the previous rural vernacular names, so the Metroplex becomes the dominate regional name. In the case of the Metroplex, it is the obvious core of North Texas. Meanwhile, Austin might be the largest city in Central Texas, but it does not define or dominate the region. El Paso statistically dominates the westernmost part of the state, but El Paso is as much borderlands than typical West Texas. Furthermore, El Paso is really only a part of the larger binational *Paso del Norte* metropolitan area. The other major cities of Houston and San Antonio are not so obvious. Houston is clearly the core city for its large metropolitan area, which also tops out at over five million residents. Houston is difficult to place in any of the major regions because its large population instantly dominates the regional data. Currently, Houston is best classified as the Gulf Coast which has popular support. As previously discussed, San Antonio is not a clear-cut case for regional association. In fact, the most distinguishing characteristics about large Texas cities are their similarities with each other. The economic engine of Texas cities commands a lion's share

FIGURE 12-15 Stunning Growth and Revitalization in McKinney
Old county courthouse used as community center and adjoining square being revamped, Collin County. Photo by Author.

The largest metropolitan area in Texas has both new and old landscapes, but it also has sharp contrasts. Deep Ellum looks like a post-industrial relic with abandoned warehouses and railroad lines. However, this corner of downtown Dallas is slowly being refurbished as an alternative/entertainment district. McKinney is the county seat for Collin County, which is growing exponentially, as it absorbs growth North of Dallas. The historic courthouse square is being refashioned into an upscale shopping and entertainment area.

of growth and wealth, and it is where the future of Texas lies. As a region, Urban Texas must be drawn non-contiguously. Each urban area is its own functional zone in terms of commuting patterns and local interactions, but they are increasingly interconnected with each other and with cities around the country and world. Those Texas cities along the Mexican border are really just half of a trans-border functional urban region. The expected population growth will continue to be focused on urban Texas.

West Texas

West Texas is the largest of the cardinal direction regions, and it is the most recent. West Texas came into being after the westward moving frontier expanded into the Great Plains. Many of the initial settlers came from North Texas (Blacklands and Cross-Timbers) and comprised mainly of Upper Southerners, but other Texans and Americans participated in this frontier expansion. Significantly, their arrival often coincided with the railroad, which differs from the river and road patterns of earlier Texas. West Texas is distinct in one other way because it was organized by the state government prior to settlement. Therefore, property lines were drawn rectangular and counties were planned and even named by the state legislature in Austin. The general geometric alignment of property and counties did not apply to the railroads and subsequent highways, which were built to connect distant places as much as the local ones.

Scholars conceptualize West Texas as a conservative Anglo area of Texas, where the acronym WASP is most applicable. In addition, West Texas has the stereotypical Texan plain environment with more ranching than farming. The majority of the land is used extensively for cattle ranching; however, intensive farming based on irrigation is quite valuable and productive. The discovery and development of the West Texas oil fields introduced wealth and outsiders that countered the trends of demographic stagnation. Currently, new arrivals, if any, are increasingly Hispanic, which is changing the demographic composition. West Texas is popularly perceived all the way to El Paso, which is larger than the other cities of Lubbock, Abilene, Wichita Falls, and San Angelo, but does not serve as a core. The number of popular vernacular regions in part or whole that correlate with West Texas is ten. Big Country and Rolling Plains are part of the Interior Lowlands, while the Southern High Plains is part of the Great Plains. Moving southward, the Permian Basin, Concho Valley, and Edwards Plateau occupy the area between Central Texas and Far West Texas. Far West Texas includes the Trans-Pecos, Big Bend, Davis Mountains, and Sun Country vernacular regions. The southern margins of West Texas overlap with the Borderlands region, and this is the second dynamic area of Tex-Mex hybrid cultural elements. Most of the actual cross border activity is in the Upper Valley region within the El Paso—Juarez urban area.

Panhandle

Somewhere between Lubbock and Amarillo the primary association with West Texas and Panhandle transitions. The Panhandle region derives from the cartographic image of the state. Twenty-five counties including Amarillo identify with the vernacular Panhandle. Academics tend to be more circumspect when they prioritize the cultural implant of Lower Midwestern culture. Therefore, academics limit the Panhandle to the northernmost counties primarily north of the Canadian Breaks. On the other hand, popular notions of the Panhandle extend southward to fit the geometric shape. IH-40 is one of the major East-West freeways crossing the United States, and the actual highway in Texas approximates the historical Route 66. In this regard, Amarillo is better connected to California and Missouri than the Mexican border or even Austin.

FIGURE 12-16 Go Further West to Marfa
Railroad designed towns are common in West Texas with linear landuse and street patterns, Presidio County. Photo by Author.

FIGURE 12-17 Adrian as midway point of Route 66
Direction and mileage pole in front of Bent Door Service Station, Oldham County.
Photo by Author.

American car culture is apparent along Route 66, and Cadillac Ranch outside Amarillo exemplifies this well. The Mother Road, made famous in *Grapes of Wrath*, was mostly displaced when Interstate 40 was constructed. Little sections of the historic road are still found in the cities and towns, where the freeway bypasses them. One such place is Adrian, where the old AAA road marker states that this is the halfway point between Chicago and Los Angeles. The Bent Door service station is closed but the Midway Restaurant still serves customers. The decor and the customers tend to reflect the cultural fascination of Route 66.

FUTURE GEOGRAPHIES

What will Texas look like in ten years or ten decades? It maybe too far away to predict for 2100, but the next decade or two will likely follow some of the prevailing trends. Population growth and demographic change will be an obvious feature of the state. People from other parts of the United States and the world will continue to migrate here. Some predictions put the population at or near thirty million by 2020. Sometime later during that same decade, Hispanic population will have nearly transitioned from plurality to majority status. As the state re-Hispanicizes, we can expect subtle changes in government priorities, the representation of Catholicism as the largest religion, and a larger percentage of everyday folk speaking Spanish and being bilingual. The reactions of Anglo-Texans will probably set the tone for how smoothly this transition occurs. A unique opportunity to construct and promote a "Texan" identity that incorporates all who live here will be short lived. Once the political system devolves along ethno-cultural cleavages, it will not be easy to recover and have a multicultural Texan identity.

Rapid population growth will spawn a host of human–environmental pressures. Foremost, the sprawling growth of Texas cities will continue to encroach on agricultural lands and natural environments. Most large cities were founded on good sites that were selected for their locational advantages such as crossroads and river fording combined with agricultural potential. As urbanization progressed, cities became larger at the expense of adjacent farms and ranches to accommodate industrial scales of production. Today, new suburban developments are irreversibly changing land use, but this is a choice and not a necessity. Urban Texas is not dense by comparison with other urban areas, and much of the existing city could be redesigned to hold more residents. Yet, the prevailing trend is to keep expanding outward. Each acre converted to urban development is an acre that will never be farmed and an acre that natural ecosystems will never be able to return. It also means a larger urban interface, which means more contact with problems such as wildfire and wild animals.

Infrastructure requirements will have to keep pace with population and settlement growth; however, the exact future form might be different because of the continual evolution of telecommunications. For now, new suburban developments have to be integrated with the city in terms of basic infrastructure such as electricity, water, sewer, and roads. Except for communication, connecting these peripheral areas is expensive, and much of the cost actually falls on the preexisting city residents. Perhaps, the biggest infrastructure question revolves around the future of automobiles. The high value we place on individual mobility that automobiles provide is likely to continue far into the future. The likely changes will be under-the-hood with different fuel combinations and electric hybrid engines, but those changes will create new delivery and supply patterns for those specific elements.

It is clear that the state will construct new highways and add lanes to existing highways, but the exact format and pattern is still to be decided. Any plan will create winners and losers. Plans such as the Trans-Texas Corridors would provide new capacity and new opportunities along the new sections. Will those areas bypassed by new roads be high-and-dry as places were when the railroad and interstate highway bypassed? Opposition to the Trans-Texas plan will come from the landowners forced to sell, communities that are not included, and those who see the investment in automobile infrastructure as unwise. Regardless, landscape change with development that accommodates automobiles is the norm and will continue into the near future.

As an exploration in regional geography, we have to ask what sorts of regions and regional names will Texas have in the future. The resiliency of historic place- names is tremendous; Texans still use native and colonial place names such as Waco and San Felipe. Words and names already in our inventory are going to survive. The regions themselves undergo change. Functional urban regions are expanding beyond city and county boundaries, and one of our most prominent examples of a recently introduced name, Metroplex, appears in this context. Rural areas with stagnant or declining populations are changing as well. With the introduction of limited regional governance, new associations and pursuit of new development opportunities may produce both real change and new names. The Councils of Governments stand out in this regard because they actively promote economic investment and tourism activities. The inflated boundaries of Alamo, Big Bend, and Big Thicket as well as the colorful Golden Triangle, Sun Country, and Heart of Texas remind us of the power of imagination. The Council of Government structure might actually evolve into new forms of governance when and if small counties decide it is in their best interest to consolidate.

FIGURE 12-18 Hurricane Ike response, Kemah
Local merchants show resiliency after the hurricane's disruption, Galveston County. Photo by Author.

Virtual geographies are the biggest unknown. It is still too early to tell if virtual communities will amount to anything other than a temporary fad. At first glance, virtual communities are ageographical because these assemblies of individuals lack proximity and known location. However, the language of communication and topical interests are just two elements that tend to ground virtual associations. Existing identities such as Texan have the opportunity to use cyberspace as another way to communicate and evolve the discourse. Texans can keep in touch through the medium; for example, military members deployed overseas can stay informed with personal events and community affairs through the Internet. Of course, it can work the other way. New migrants to Texas can stay informed and engaged in their previous communities. Therefore, we might ask if earthly boundaries really mean anything if one can both choose where to be and who to be? The fact that we have the liberty to ask and the learning to pose that question answers it.

CONCLUSION

Where is Texas? Who are Texans? What are places like in Texas? How should we describe Texan identity, territory, and nature? So far our geographical inquiry has asked these questions. We began our inquiry with some initial definitions and parameters of both Texas and geography. Our journey took us through the cultural diversity and historical evolution of Texas. Then, we traveled over the literal earth as we discussed the physical geography of Texas. We continued on by looking at the contemporary human geography of the state. We just elaborated our discussion by returning to the concept of regions and constructing the major regions of Texas. Finally, we concluded with a brief foray into the future and speculate about the trends.

The geography of Texas is probably better called the geographies of Texas. With so much space and so many different cultures, Texas is fractured along both human and physical criteria. Despite some serious stress points, Texas is not broke. It is strong and vibrant; Texas continues to attract new migrants, new capital, and new imagination. In a strange manner, Texas is both one and many. After all this discussion we return to that key question and settle on this reality. Texas is one entity that does have a common framework. Past geographies followed a path that produced a singularly quilted Texas with mixed backgrounds and contributions. The geographies that illuminate structure and organization point to more than a state of mind and to a grounded state of earthly jurisdiction. On the other hip, an equally strong argument for pluralism and diversity can be made that confounds any single imagination of a Texas stereotype. Our only hope is to embrace regions and accept Texanity in all its guises, manifestations, and landscapes. Geographies of the people, places, and patterns reveal Texians, Texases, and Texanities.

Further Readings

Daniel D. Arreola
    "The Mexican American Cultural Capital," *Geographical Review,* Vol. 77, No. 1, pp. 17–34, 1987.
    Tejano South Texas: A Mexican American Cultural Province. Austin: University of Texas Press, 2002.

Richard Francaviglia
    The Shape of Texas: Maps as Metaphors. College Station: Texas A&M University Press, 1995.

Pete Gunter and Max Oelschlaeger.
    Texas Land Ethics. Austin: University of Texas Press, 1997.

Historical Atlas of Texas. Stephens and Holmes, Oklahoma University Press, 1989.

John Brinckerhoff Jackson
    "Chihuahua as we might have been," *Landscape,* Vol. 1, No. 1/Spring, pp. 16–24, 1951.
    "High Plains," *Landscape,* Vol. 3, No. 3/Spring, pp. 1–3, 1954.
    The Southern Landscape Tradition in Texas. Amon Carter Museum, Fort Worth, 1980.
    "The Vernacular City," *Center,* Vol. 1, pp. 27–43, 1985.

Terry G. Jordan (Jordan-Bychov)
    "The Origin of Anglo-American Cattle Ranching in Texas: A Documentation of Diffusion from the Lower South," *Economic Geography,* Vol. 45, No. 1, pp. 63–87, 1969.
    "Population Origins of Texas, 1850" *Geographical Review,* Vol. 59, No. 1, pp. 83–103, 1969.
    "Early Northeast Texas and the Evolution of Western Ranching," *Annals of the Association of American Geographers,* Vol. 67, No. 1, pp. 66–87, 1978.
    "Perceptual Regions of Texas," *Geographical Review,* Vol. 68, No. 3, pp. 293–307, 1978.
    Texas Graveyards: A Cultural Legacy. Austin: University of Texas Press, 1982.

Terry G. Jordan (with J. Bean & W. Holmes)
    Texas: A Geography. Boulder: Westview Press, 1984.

Donald W. Mcinig

Imperial Texas: An interpretive essay in cultural geography. Austin: University of Texas Press, 1969.

Southwest: Three Peoples in Geographical Change, 1600–1970. Oxford University Press, 1971.

Char Miller and Heywood T. Sanders, editors

Urban Texas: politics and development. College Station: Texas A&M University Press, 1990.

Richard L. Nostrand

"The Hispanic-American Borderland: Delimitation of an American Culture Region," *Annals of the Association of American Geographers,* Vol. 60, No. 4, pp. 638–661, 1970.

R. J. Russell

"Climates of Texas," *Annals of the Association of American Geographers,* Vol. 35: Issue 2/June, pp. 35–52, 1945.

Darwin Spearing

Roadside Geology of Texas. Missoula: Mountain Press Publishing Company, 2002.

Eric R. Swanson

Geo-Texas: A Guide to the Earth Sciences. College Station: Texas A&M University Press, 1995.

Texas Almanac. 2006–2007/63rd edition. Dallas Morning News and Texas A&M University Press.

U.S. Census Bureau, *2000 Census*

Congressional Apportionment

Population Change

Texas Abstract

Robert E. Veselka

The Courthouse Square in Texas. Austin: University of Texas Press, 2000.

Harry Walsh and Victor Mote

"A Texas Dialect Feature: Origins and Distribution," *American Speech,* Vol. 49, No. 1/2, pp. 40–53, 1974.

Writers' Program (Texas)

Texas: a Guide to the Lone Star State.

Wilbur Zelinsky

"North America's Vernacular Regions," *Annals of the Association of American Geographers,* Vol. 70, No. 1, pp. 1–16, 1980.

List of Maps

Figure A   Location Map Reference   3

Figure B   Location Map: Cities and Rivers   4

Figure C   Location Map: Cities and Counties   5

Figure D   Tejas blank map   6

Figure 1-2   State of Texas   10

Figure 1-4   American Regions   14

Figure 1-5   Texas Neighbors   15

Figure 1-6   Texas Border Segments   18

Figure 3-2   Mexican Texas: Castenada's map of Texas, 1820–1836   44

Figure 3-3   1492 Americas   47

Figure 3-4   Native Texas   49

Figure 3-9   Republic of Texas   58

Figure 4-2   Diffusion of American Subcultures   65

Figure 4-3   European Texas   67

Figure 4-5   Language Regions in Texas   74

Figure 4-7   Religious Regions in Texas   77

Figure 5-8   Historic San Antonio   91

Figure 5-10   Historic Town Squares   93

Figure 5-11   Galveston City Plan   94

Figure 5-12   Austin City Plan   95

Figure 5-13   Laredo/Nuevo Laredo   96

Figure 5-15   Folk Architecture Regions in Texas   99

Figure 5-22   Texas Ranching Regions   106

Figure 6-2   NOAA Relief Image of Texas   114

Figure 6-4   Texas Geological Regions   119
Figure 6-6   Average Annual Temperature   122
Figure 6-7   Average Annual Precipitation   123
Figure 6-9   Climatic Classification of Texas   127
Figure 6-11   Major River Basins of Texas   129
Figure 6-13   Major Aquifers under Texas   132
Figure 6-14   Vegetation Regions of Texas   133
Figure 6-18   Land Resources of Texas   137
Figure 7-2   North American Physiography   143
Figure 7-3   Intermontane Physiography   145
Figure 7-7   Great Plains Physiography   148
Figure 7-13   Interior Lowlands Physiography   153
Figure 7-16   Coastal Plain Physiography   156
Figure 7-21   Upper Coast of Texas   161
Figure 7-23   Coastal Bend of Texas   162
Figure 7-25   Embayment of Texas   164
Figure 7-28   Gulf of Mexico   166
Figure 8-2   Agricultural Regions   171
Figure 8-5   Agriculture: Cattle and Cotton   173
Figure 8-6   Texas Reservoirs   178
Figure 8-7   Oil and Gas Fields   179
Figure 9-2   Metropolitan Statistical Areas   192
Figure 9-3   Population Distribution   196
Figure 9-5   Minority & Majority Texas   201
Figure 9-7   Urban Clustering   206
Figure 10-2   1861 Secession Vote   217
Figure 10-3   Texas Political Regions   219
Figure 10-4   Electoral Geography: Partisan Voting Behavior   220
Figure 10-5   Electoral Geography: Recent Statewide Elections   221
Figure 10-6   Congressional Districts   223
Figure 10-7   Council of Governments   224
Figure 10-10   Employment   227
Figure 10-11   Income   228

Figure 10-12   Economic Regions   230

Figure 10-13   Transportation Infrastructure   233

Figure 11-7   Tourist Regions   248

Figure 11-19   Texas Touring Trails   256

Figure 12-3   Cardinal Direction Regions   271

Figure 12-4   Perceptual Regions of Texas   272

Figure 12-7   Meinig's Regions of Texas   276

Figure 12-8   Jordan's Traditional Rural Texas Regions   277

Figure 12-10   Major Regions of Texas   280

Index

A

absolute location, 32
academic geography, 28, *29*
academic synthesis, 275–277
Adams–Oñis Treaty, 23, 56
administrative districts, 222–225
Adrian, *289*
African-American (Black) migration, 71
African-American Vernacular English (AAVE), 73
age and gender structure, 198
agriculture, 170, 181
    cattle-driving, *173*, 174
    cotton-picking, 172, *173*
    crops, 174
    data, *172*
    livestock, 174
    regions, *171*
air mass boundaries, 120
airports, 234
Alpine, *288*
alternative energy, 181
American archaeology, 46
American Regions, *14*
amusement parks, 258–259
Anahuac, 57
Anglo-American (White) migration, 64–66
annual temperatures, average, *122*
anthropogenic hazards, 186–187
Appalachian mountains, 116, 142
aquiclude, 130
aquifer, 128, 130, *132*, 176, *177*, 178
Aransas Bay, 159
Aransas National Wildlife Refuge, 159
Association of American Geographers (AAG), 28
Atlantic and Gulf Coastal Plain, 141, 142
atmosphere, 115
Austin, 207, *211*, 279, 286
    city plan, *95*
Austin, Moses, 56
Austin, Stephen F., 56
Aztec Empire, 51

B

Baffin Bay, 162
Balcones Escarpment, 149, 154
Balcones Fault Zone (BFZ), 116, 149
Balconies Escarpment, 145
ballot box vote, 216
Balmorhea Springs, 134
bars and coffee shops, 262
Bay of Campeche, 163
Beaumont-Port Arthur MSA, 207
Berkeley school of geography, 83
Bexar, 57
Bible Belt, 29, 77, 86
Big Bend, 142, 144, 250–251, 275
Big Country, 275
Big Lake, 176
Big Thicket, 157, 247, 275
biodiversity, 174–175
biogeographical processes
    soils, 138–139
    vegetation, 130–134
    wildlife, 134–138
biosphere, 115
Blackland Prairies, 155, *158*
blue-northers, 120
Border English, 74
Borderlands, 283–285
borders and neighbors, 13–24, *17*, *18*
    border posts, *21*
    length, *19*
    Texas comparisons, *16*
Boundary Commission (1884), 19
Brazoria, 59
Brazos Island, 162
Brownsville-Harlingen MSA, 207
Brush Country, 133–134, 157, *160*, 251–252, 272

C

Caddo, 50
Caddo Lake, 176
Cajuns, French-speaking, 69
Canadian Breaks, 147
Canyonlands, 149
Caprock Escarpment, 145
cars, love for, 241
Cartesian coordinates, 38–39
cartograms, 38
cartographic scale, 33
cartography, 34, 35
Catholic cemeteries, 88
cattle-driving, *173*, 174
cemeteries, 85–86
    Mexican and German, 87–88
    modern, 88, 90
    southern folk, 86–87
Cenozoic materials, 117
Central Business Districts (CBDs), 210, *212*
Central Texas, 279
    German Hill Country, 282
    Gulf Coast, 282
Central Texas Uplift, 150

Cherokee, 51
Chihuahua, 13, 17
Chisos Mountains, 144
choropleth maps, 38
cities, 202–204
    largest, *208*
citification, 201
city plans, 92–97
civitas, 201–202
    Metropolitan Statistical Areas (MSA), 204–213
    Texas cities, 202–204
cleaning days, 87
climatic hazards, 185–186
climatic processes, 117–118
    classification, 123–126, *127*
    global factors, 118–121
    precipitation, 121–122, *123*
    temperature, 121
climographs, *124–125*
Clovis, 45–46
Coahuila, 13, 17, 57
Coahuiltecan, 50
Coastal Bend, 158, *162*, 274
    Aransas Bay, 159
    Corpus Christi Bay, 159
    Matagorda, 158
    San Antonio Bay, 159
coastal flooding, 185
coastal marsh, 134, *161*
Coastal Prairies, 157
Colorado River, 158
Columbian Exchange, 46–48
Comanche, 51, 153
Compromise of 1850, 61
congressional districts, *223*
conservation, 181–182
conservation tourism, 245–246
continentiality, 120
Corn Belt, 29
Corpus Christi Bay, 159
    Waterfront, *167*
Corpus Christi MSA, 207
Cotton Belt, 29, 30
cotton-picking, 172, *173*
Council Of Governments, 224–225
counties, largest, *209*
county courthouse model, *101*
courthouse squares, 102–106, *103*, *104*
cratons, 116

crops, 174
Crosby, Alfred, 46
Cross Timbers, 108, 134, 153–154, *155*, 272
cruising, 263
cultural diversity, 63
    defining culture, 63–64
    languages, 72–76
    migrant origins, 64–71
    religions, 76–81
cultural hearth, 48, 51, 65, 108
cultural historical geography, 41–43
cultural landscapes, 83–85
    cemeteries, 85–90
    courthouse squares, 102–106
    folk architecture, 97–100, 102
    land and organization, 90–97
    ranching, 106–108
culture, defining, 63–64
Czech communities, 68

D

Dallas, 203, 286
Dallas-Fort Worth MSA, 206
dance halls and live music, 261–262
Davis Mountains, 144
definition, of Texas
    essential, 7–9
    locational, 9–13
De Leon, 56
demography. *See* population
development, 226–227
    economic regions, 229–231, *230*, *231*
    quality of life, 227–230
De Witt, 56
diffusion, 42
    of American subcultures, *65*
    relocation, 81
diversion tourism, 246–247
drainage basins, 128
drought, 185
dynamic quality, of regions, 32

E

earthquakes, 182–183
earth system, 115, *117*
earth writing, 27–29
Eastern Cross Timbers, 154
East Texas, 285–286

East Texas Timberlands, 156–157
economic geography, 225–226
    development, 226–232
    infrastructure, 232–236
economic indicators, *229*
economic regions, 229–232, *230*, *231*
education and information, 235–236
Edwards Plateau, 116, 145, 148–149, 183, 282
    Canyonlands, 149
    Lampasas Cut Plain, 149
El Camino Real, 52
election signs, *225–226*
electoral behavior, 216–218
electoral districts, 218–222, *220*, *221*
elevation, 119–120
El Paso, 11, 19, 52, 121, 122, 142, 248, 286
    MSA, 207
employment, *227*
Empresario system, 56
Enchanted Rock, 117, 150, *152*
enclosure, 163
environmentalism, 187
European (White) migration, direct, 66–69
European Texas, *67*
expansion diffusion, 42
external relations and regions, 32
extra-tropical storms, 185

F

fences, 108
    wood, 98
floods, 185
Flower Garden Bank National Marine Monument, 164
Flower Gardens, 164
fluvial flooding, 185
folk architecture, 97–100, 102
    regions, *99*
foreign-born population, *69*
formal regions, 30
Fort Worth, 203, 286
fossil fuels, 180–181
Francaviglia, Richard
    The Shape of Texas, 24
Franklin Mountains, 144
Fredericksburg, 100, 145

Free State, 273
fronts, 120
functional regions, 30–31

G

Galveston, 33, 202, 204, 207
    Bay, 158, 203
    beach, *253*
    city plan, *94*
general purpose maps, 35
geography, definition of, 27
geographical past, 45
    Clovis, 45–46
    Columbian Exchange, 46–48
    Mexican Texas, 56–60
    Native Texas, 50–51
    Republican Texas, 60
    Spanish Texas, 52–55
    statehood, 61
geographies of play, 256, 258
    Kids, 258–260
    nightlife, 265–267
geologic hazards, 182–183
geologic processes, 116
    profile, *120*
    regions, *119*
    S-curve line, 116
    surface expressions, 116–117
    time factor, *118*
geo-spatial techniques, 28
geothermal electricity, 181
German-Americans, 67–68, 75
German cemeteries, 88
German Hill Country, 282
German Ten, 282
globalization, 226
global warming, 186–187
global water percentages, *126*
Golden Triangle, 274
Goliad, 55
Grand Prairie, 153
Greater Houston Region. *See* Houston-Baytown-Sugarland
Great Plains, 141–142, 145
    Central Texas Uplift, 150
    Edwards Plateau, 148–149
    High Plains, 145, 147
    physiography, *148*
    Stockton Plateau, 147–148
    Toyah Basin, 147–148
Green Lake, 176–177

Greenwich Naval Observatory, 39
ground water, 129–130, 177–178
Guadalupe Range, 144
Gulf Coastal Plain, 154, 252, 282, 286
    Blackland Prairies, 155
    Coastal Bend, 158–159
    Coastal Prairies, 157
    Interior Coastal Plain, 156–157
    physiography, *156*
    Rio Grande Embayment, 162–163
    Upper Coast, 157–158
Gulf Coast Aquifer, 178
Gulf of Mexico, 163–165, *166*
    profile, *167*
Gulf Stream, 165
Gunter, Pete A. Y.
    Texas Land Ethics, 187
Gypsum Plains, 151

H

Heart of Texas, 275, 279
High Plains, 145, 147
Hill Country, 149, 252–253
historical infrastructure, *235*
Historical Square, *249*
historical tourism, 243–245
horse-shoe lakes, 176
Horst and Grabens, 141
Houston, 203, 207, 286
    ship channel, *211*
Houston-Baytown-Sugarland, 207
Hueco Tanks, 134
human environment interactions, 169–170
    agriculture, 170–174
    biodiversity, 174–175
    environmentalism, 187
    natural hazards, 182–187
    natural resources, 175–182
human geography, 28
    modern, 189–190
human mobility, 42
hurricanes, 185, *186*
hydroelectricity, 181
hydrological cycle, 126, *124–128*
hydrologic processes, 124–128
    ground water, 129–130
    surface water, 128–129
hydrosphere, 115

I

ideological politics, 216
IMAX Theater, *270*
income, *228*
infrastructure, 232
    education and information, 235–236
    electricity, *234*
    historical, *235*
    ports and airports, 233–234
    railroads, 232
    roads, 232
    transportation, *233*
    utilities, 234–235
Interior Coastal Plain, 156
    Brush Country, 157
    East Texas Timberlands, 156–157
    Post Oak Savannas, 157
Interior Lowlands, 141, 142, 151
    Cross Timbers, 153–154
    Grand Prairie, 153
    Osage Plains, 151
    physiography, *153*
Intermontane Basins and Plateaus, 141, 142
    Basin and Range Province, 144
    physiography, *145*
internal relations and regions, 32
International Boundary and Water Commission, 19
involuntary migration, 42
Irish Empresarios, 56
Irregular Rectangles survey system, 90
isopleth maps, 38

J

Jackson, J.B., 15, 17
Jefferson, 285
Jordan, Terry, 217, 270, 275, 276, *277*, 279
    Texas: a Geography, 8

K

Karankawas, 50
Kids' play, 258
    amusement parks, 258–259
    movie theatres, 260
    playgrounds and parks, 260

Kids' play (*continued*)
  rodeos, fairs, and festivals, 259–260
  shopping malls, 259
  special events, 260
Killeen-Temple MSA, 207
Köppen system, 126

L

La Bahia, 55, 57, 282
Laguna Madre, 162
Lake o' the Pines, 176
Lampasas Cut Plain, 149
land and organization, 90
  city plans, 92–97
  survey systems, 90–91
land ethics, 187
land resources, *137*
landscapes. *See* cultural landscapes
landslides, 183
land use, 181
languages
  English, 72–73
  regions, *74*
  Spanish, 73–75
Laredo, *96*, 284
LaSalle, 52
latitude, *37*, *38*, 38–39, 118
Lavaca River, 158
Law of the Indies (1573), 92
leisure and play, 239–241
  geographies of play, 256, 258–263
  tourism regions, *245*, 247–255
  travel and tourism, 242–247
Leopold, Aldo, 187
lithosphere, 115
livestock, 174
Llano Estacado, 147
Llano river, *131*
location, 32
log cabins, 97–98
longitude, *37*, *38*, 38–39
Long Lots survey system, 90
Loop Current, 165
Los Adeas, 202
Lost Maples, 134
Lost Pines, 134
Louisiana, 13
Luckenbach, 281

M

maps, 33–34, *271*, *272*, 275
  in Alley, *278*
  reading, 34–35
  trails, *256*
  types of, 35–38
Maquiladoras, 144, 230
Marathon Uplift, 144
maritime polar air masses (mP), 120
maritime tropical air mass (mT), 120
Matagorda, 158, *163*
Mayan Empire, 51
McAllen-Edinburg-Pharr MSA, 207
McKinney, 287
McMullen County, 273
Mediterranean Sea, 163, 165
Meinig, Donald, 275, *276*, 279, 286
Mesoamerica, 51
Mesozoic aged materials, 117
Metes and Bounds survey system, 90
Metroplex, 273–274, 286
Metropolitan Statistical Areas, *192*, 204, 206, 286
  metropolitan landscapes, 207–213, *210*
  Texas, 206–207
Mexican-American (Hispanic) migration, 69–71
Mexican and German cemeteries, 87–88
Mexican Texas, *44*, 56–60
Mexico, 13, 16, 19
mid-Atlantic folk architecture, 99
migrant origins
  African-American (Black), 71
  Anglo-American (White), 64–66
  direct European (White), 66–69
  Mexican-American (Hispanic), 69–71
migration, 42, 68, 69–71
"Minority-Majority" situation, 200, *201*
Mississippi Fan, 164
Mississippi River, 142
modern, 190
modern cemeteries, 88, 90
modern city model, *205*
modernity, 190
movie theatres, 260
multilingual signs, *75*
municipalities, 57
museums, 244

N

Nacogdoches, 55, 57
National Oceanic and Atmospheric Agency relief images, *114*
National Park System, 182
National Wildlife Refuges, 182
native Texas, *49*, 50–51
natural hazards, 182
  anthropogenic, 186–187
  climatic, 185–186
  geologic, 182–183
natural lake, 176
natural resources, 175
  conservation/recreation, 181–182
  land use, 181
  oil, 179–181
  water, 176–178
Neches River, 176
necrogeography, 85–86
neighbours and borders, 13–24
New Forest, *249*
New Mexico, 13
New Orleans, Louisiana, 33
nightlife, 261
  bars and coffee shops, 262
  body appearance, 262–263
  cruising, 263
  dance halls and live music, 261–262
  strip clubs and brothels, 262
North of the Breaks, 147
North Texas, 286
  Urban Texas, 286–287
Nueces Plain, 157
Nuevo Laredo, *96*
Nuevo León, 11, 17

O

oaklands, 134
Oelschlaeger, Max
  Texas Land Ethics, 187
Ogallala Aquifer, 130, 178
oil
  alternative energy, 181
  field equipment, *180*
  fossil fuels, 180–181
  and gas fields, *179*
Old World, 48
Osage Plains, 151
  West Texas Rolling Plains, 151
Ouachita Mountain Range, 116
"Out-of-Africa" thesis, 48

P

Padre Island, 162
Paleozoic materials, 117
Palo Pinto, 154
Panhandle Plains, 251, 273, 288–289
partisan politics, 216, 219, *220*
past rank orders, 202–203
Patrón system, 218
Pecos River, 148, *151*
perceptual regions, 270–275
   cardinal direction regions map, 271
   names, *273*
   of Texas map, *272*
Permian Basin, 116, 272
physical geographies, 28, 111–113
physical processes, 115–116
   biogeographical processes, 130–139
   climatic processes, 117–126
   geologic processes, 116–117
   hydrologic processes, 126–130
physiographic regions, 141, *143*
   Great Plains, 145–151
   Gulf Coastal Plain, 154–163
   Gulf of Mexico, 163–166
   Interior Lowlands, 151–154
   Intermontane Basins and Plateaus, 142–145
   North American provinces, 141–142
physiography, 112
Piney Woods, 132–133, 157, *159*, 249, 285
place, 238
placeless, 238
place-utility theory, 42
Plains Indians, 50–51, 61
playa lakes, 176
playgrounds and parks, 260
Poles, 68
political districts, 218
   administrative, 222–225
   electoral, 218–222, *220*, *221*
political geography, 216
   electoral behavior, 216–218
   political districts, 218–222
political regions, 218, *219*
Polk, James, 61
pollution, 186
population, 193
   change, 194
   density, 195
   distribution, 195–198, *196–197*
   characteristics, 198–201, *199*
   composition, *200*
Port Arthur, 121, 286
Port Lavaca, 158
ports, 233–234
post-modern, 190
Post Oak Savannas, 157, *160*
poultry, 174
prairies
   and lakes, 255
   and steppe grasslands, 134
Precambrian rock, 117, 150
precipitation, 121–122, *123*
primary economic activities, 225–226
Prime Meridian, 39

Q

quality of life, 227–229
Quaternary materials, 117

R

racial and ethnic composition, 199
railroads, 232
ranching, *105*, 106–108
   regions, *106*
rate of natural increase (RNI), 194
Raza Unida, 218
Reconquesta, of Iberian Peninsula, 92
recreation, 182
Red Beds, 151
Red River Rolling Plain, 151
regional geographies, 237–238. *See also* regions
regions, 269
   characteristics of, 31–32
   definition of, 30
   future geographies, 289–291
   latitude and longitude, 38–39
   location and scale, 32–33
   major, 277–289, *280*
   maps, 33–38
   popular, 269–277
   types of, 30–31, *31*
Regular Rectangles survey system, 90–91
relative location, 32
religions, 76–81
   regions, *77*
relocation diffusion, 42
Republican Texas, 60
resacas, 176
reservoirs, *176*, 177, *178*
Richardson, Miles, 64
Rio Grande, 11, 15, 17, 19, 61, 144, 274
Rio Grande Embayment, 162, *164*
   Lower Valley, 163
rivers, *130*
   basins, *129*
   flooding, 185
roads, 232
Rocky Mountains, 141, 145
rodeos, fairs, and festivals, 263–264
Roosevelt, Teddy, 217
Russell, R. J., 126

S

Sabine Lake, 176
Sabine Uplift, 157
Said, Edward
   Orientalism, 64
San Antonio, 52, 55, *91*, *92*, 202, 207, *213*, *242*, *243*, 251, 282, 286
   Bay, 159
Sandy Hills, 157
San Felipe, 57
Sauer, Carl, 83
scale, 32–33
Scandinavian migration, 68
scraping, 87
S-curve line, 116
seasonality, 122
secession vote, *217*
secondary economic activities, 226
sense of place, 238
Sequence Occupancy, 84
settlement geography, 84–85
shape, of Texas, 24–25
shopping malls, 259
shrink-swell soils, 183
Sierra Madres, 141
Sigsbee Abyssal Plain, 164
sinkholes, 183
site, 33
situation, 33
Slovaks, 68
social structures, 215–216
   economic, 225–236
   political, 216–225

soils, 138–139
Solitario, 144
Sorbs, 68
southern folk cemeteries, 86–87
South of the Breaks, 147
South Padre, *165*
South Texas, 282
    Borderlands, 283–285
South Texas Plains. *See* Brush
    Country
South Texas Sand Sheet, 157
Spanglish, 74
Spanish city model, *53*, 92
Spanish herding, 107
Spanish Louisiana, 55
Spanish Texas, 52–55
special events, 260
Spindletop, 180
Standard American English
    (SAE), 76
statehood, 61
Stockton Plateau, 147–148
storms hurricanes, 185
Straits of Florida, 163
strip clubs and brothels, 262
strom tracks, 120
subsidence, 183
surface expressions, 116–117
surface water, 128–129, 176–177
survey systems, *89*, 90–91

T

Tamaulipas, 17
Tejano, 74
Tejas, 282
temperature, 121
tertiary economic activities, 226
Texas cities, 202
    past rank orders, 202–203
    urbanization, 204
Texas Heritage Trails Program
    (THTP), 255
Texas-is-Many group, 8
Texas-is-One group, 8

Texas-Louisiana Shelf, 164
Texas-Louisiana Slope, 164
Texas Revolution, 60
Texas Urban Triangle, 207
Texas Vernacular English (TVE), 76
Texoma, 273
thematic maps, 38
    types, *36*
Thurmond, Strom, 218
tidal surges, 165
Tigua, 51
Tonkawa, 50
topographic maps, 35, 38
tornadoes, 185
tourism regions, 247, *248*
    Big Bend Country, 248–249
    Gulf Coast, 252
    Hill Country, 252–253
    Panhandle Plains, 251
    Piney Woods, 247
    prairies and lakes, 255
    South Texas Plains, 251–252
Town Squares, *93*
Toyah Basin, 147–148
Trans-Pecos, 134, 142
transportation infrastructure, *233*
travel and tourism, 241
    conservation tourism, 245–246
    diversion tourism, 246–247
    historical tourism, 243–245
Treaty of Guadalupe-Hildago, 61
tropical cyclones, 185
typhoons, 185

U

uniform regions. *See* formal
    regions
Upper Coast, 157–158
    Galveston Bay, 158
Upper Valley, 144
urban clustering, *206*
urban heat island, 170
urbanization, 195, 204
Urban Texas, 286–287

U.S.–Mexico Border Commission,
    163
utilities, 234–235

V

Van Zandt County, 273
vaqueros, 107
vegetation, 130–132
    anomalies, 134
    Brush Country, 133–134
    coastal marsh, 134
    far west, 134
    oaklands, 134
    Piney Woods, 132–133
    prairies and steppe grasslands, 134
    regions, *133*
vernacular regions, 31, 270
virtual geographies, 291
voluntary migration, 42

W

Wallace, George, 218
water, 176
    and electricity infrastructure, *234*
    ground water, 177–178
    reservoirs, 177
    surface water, 176–177
westerlies, 120
Western Cross Timbers, 154
West Florida Shelf, 164
West Texas, 287–288
    Panhandle, 288–289
Wichita, 51
wildlife, 134–138
World Travel and Tourism
    Council, 242

Y

Yucatan Channel, 163
Yucatan Shelf, 164

Z

Zócalo, 201